it's a matter of

SURVIVAL

it's a matter of SURVIVAL

NEW ENGLAND INSTITUTE
OF TECHNOLOGY
LEARNING RESOURCES CENTER

anita gordon
david suzuki

Harvard University Press
Cambridge, Massachusetts
1991

Copyright © 1990 by Anita Gordon and David Suzuki
All rights reserved
Printed in the United States of America
10 9 8 7 6 5 4 3 2 1

This book is printed on acid-free, recycled paper.

Library of Congress Cataloging-in-Publication Data

Gordon, Anita.
   It's a matter of survival / Anita Gordon, David Suzuki.
      p.   cm.
   Includes bibliographical references and index.
   ISBN 0-674-46970-4
   1. Man—Influence on nature. 2. Human ecology. 3. Environmental policy.
   4. Twenty-first century—Forecasts.   I. Suzuki, David T., 1936–
   II. Title.
GF75.G66   1991
363.7—dc20   90-71337
                                                          CIP

For my father,
Jacob Gordon,
who has inspired me
with his courage and dignity,
and who taught me never to give up

— A.G.

For the two elders in my life,
Carr Suzuki and Harry Cullis,
who have been important
in keeping me on the right track

— D.S.

CONTENTS

A C K N O W L E D G M E N T S

They say that writing is a solitary endeavor; one sits for too many hours in front of a blank screen or a typewriter that does nothing but wait for something to be said. But this book is not a solitary accomplishment. The people named here stand in testament to that. Some of them you will meet in the pages that follow — scientists who explain their research clearly and their concerns eloquently. Others have stolen time from busy schedules to read portions of the manuscript for us — commenting, correcting, suggesting: thank you, Stephen Schneider, at the National Center for Atmospheric Research in Boulder, Colorado; Edward O. Wilson, at Harvard University; Herman Daly, at the World Bank; Paul Ehrlich, at Stanford University.

This book grew out of a radio series broadcast in 1989 on the national network of the Canadian Broadcasting Corporation. Nowhere else in the world would a broadcaster have taken on the subject matter in such a forthright way. The CBC is a unique broadcasting system, with unique people at its helm: people like Donna Logan and Alex Frame, who encouraged us every step of the way through both the radio series and the book.

It is said that patience is a virtue. If this is true, then Angel Guerra and Alison Reid at Stoddart Publishing are truly virtuous.

Fred Pearce in London, England, author of *Turning Up the Heat*, shared his expertise with us, and Lynn Glazier's sharp mind and sense of commitment to both the radio series and the book were greatly appreciated.

Jacob Gordon and Zev (William) Gordon went beyond the call of family and duty by not just reading portions of the manuscript, but reading, and rereading, and then rereading again, each of them critiquing from the vantage point of more than 80 years on the Earth. Stella Gordon was there through thick and thin.

And Vanessa, we're sorry we dragged your mother, Penny Park, away for all those hours she devoted to the radio series and to doing research for the book — day, night, and weekends. She did it because she believes in the message and wants to make this a better world for you.

This book could never have been written without Wilf Fielding's inspiration and wisdom.

And finally: if you look up "godsend" in an Oxford dictionary you'll find the following: "n. a piece of unexpected good luck having a decisive effect, a useful or effective acquisition." It should read: "Beverley Beetham Endersby, Editor." Thank you, Beverley, for seeing the vision, and keeping it true.

INTRODUCTION

More than any other time in history, the 1990s will be a turning point for human civilization. Not only are we facing ecological disasters that could affect our ability to survive, but the crisis is forcing us to reexamine the value system that has governed our lives for at least the past 2000 years.

The assumptions that we've made about how the natural world operates and what our relationship is to it are no longer tenable. These *sacred truths* that we've grown up with — "nature is infinite"; "growth is progress"; "science and technology will solve our problems"; "all of nature is at our disposal"; "we can manage the planet" — offer no comfort as we enter the last decade of this century. In fact, we're being told that to continue to subscribe to these assumptions is to ensure the destruction of civilization as we know it.

This book is about us as a species: our shortsightedness, our failure to read the warning signs, our inability to grasp the significance of our actions, and about the tough decisions we must make in order to save ourselves. It has grown out of a major five-hour, five-part radio series prepared for the Canadian Broadcasting Corporation. Working on the series drove home for both of us the enormity and imminence of planetary eco-collapse. In conducting more than 100 interviews with scientists and experts around the world, we were able to see the scope of the crisis and the virtual unanimity that exists among them as to the consequences of the ecological destruction that our stubborn refusal to see beyond our comfortable truths has unleashed upon the planet.

In the end, what gave the radio series its impact was the imperative formed through those interviews to look without flinching or qualification at where we are and where we are going. Never before have so many scientists spoken with a single voice, warning, as ecologist Paul Ehrlich does, "that something is seriously wrong . . . and if we wait until the full-scale catastrophe, we may or may not survive it."

The response to the programs was immediate and massive. More than 13 000 listeners wrote to us, most of them asking "What can I do?" We hope that this book will answer that question, and will act as a guide through this decade of survival.

It has become crystal clear that the planet is losing a battle with the deadliest predator in the history of life on Earth. And now, almost every day, newspapers, magazines, and radio and television news provide isolated reports that simply add more evidence for the eco-catastrophe that is under way. Once you look at events from a global ecological perspective, then a chemical spill, radioactive leak, clear-cut forest, acidified lake are all parts of the bigger picture. That's what we've attempted to portray.

The litany of environmental horrors is now a familiar one:

— the destruction of 40 hectares (100 acres) of tropical rain forest every minute, and, with it, the extinction of at least 20 000 species a year;

— the atmospheric degradation: acid rain that is sterilizing lakes and forests; the destruction of the ozone layer; carbon dioxide and methane loading that is causing global warming;
— the accumulation of massive amounts of toxic chemicals in air, water, soil, and food;
— the annual loss of billions of tons of agricultural topsoil and the consequent steady decline in total food production since 1984;
— the explosive increase in human population that is adding a quarter of a million people daily.

We tend to view each of these topics in isolation, as a separate issue, and to look for solutions in a piecemeal way. What we fail to understand is that this global crisis is a series of brush fires: while we stamp them out in one place, they flare up somewhere else. And we are the incendiary agents. We are at the center of all these issues — there are too many of us; we consume too much; we pollute too much; and we are blinded by our complacent acceptance of a dangerously outmoded system of beliefs and values.

Research groups such as the Worldwatch Institute, in Washington, D.C., tell us that we have fewer than 10 years to turn things around or "civilization as we know it will cease to exist." The simple truth is that we are the last generation on Earth that can save the planet.

In this book, we hope to convince you, as we were convinced by the portrait of this world painted by those whose vocation it is to measure and understand its complexity and unity, that we must all undergo a change in attitude if we are to survive into the next century. The psychologist Leon Festinger coined the term "cognitive dissonance" to describe the discomfort or conflict that ensues when existing beliefs or assumptions are challenged or contradicted by hard evidence.

He noticed that the mind intervenes to ease the conflict by denying the importance of the contradiction, reconciling the discrepancies, or refusing to be convinced and demanding more and more evidence. When the issue is extinction, such denial can be deadly.

In 1989, the radio series sounded an alarm that our survival is at stake. This book sounds another kind of alarm. We are being lulled into believing a new *sacred truth* — that we will be able to accommodate the salvation of the planet within our current lifestyle, that there is a quick fix for the state of the Earth that can be implemented with only slight alterations to our lives. If only it were so.

U.S. Senator Al Gore sees our challenge as "creating in a single generation a future in which people think and behave so differently that they look back on 1990, at the kind of pollution we have, at the kind of destruction under way, at the suffering we tolerate, and they wonder, as they shake their heads, how people could have thought in ways that allowed them to condone that kind of activity."

We ask you, that generation, to look back on 1990, and to visit the future. We offer no apologies for what you will be shown. The vision is bleak, certainly. The message to be read from it is not. By seeing what lies ahead we can alter our course and avoid catastrophe. In facing the reality of what we are up against, and acting to change it — and to change ourselves — there is an opportunity for hope, for faith. Children belong to all of us; we must act together to ensure them a future. In this way, this matter of survival is a matter of the heart.

PART I
TOWARD THE
YEAR 2040

ONE

BEYOND YOUR WORST NIGHTMARE

A.D. 2040. If we were to give this year a name it would be Despair. This is the hopeless world that we have left our children and grandchildren. Where once our lives were measured and enriched by the cycle of the seasons, in summer there is now only searing heat and the certainty that it will get hotter. Seasons exist only in the nostalgic longing of those of us old enough to remember the richness of life. Summer once took us to the world's coasts for the pleasure of sea and sand and cooling breezes; now, rising sea levels have driven millions of coastal and island peoples inland. Food is scarce. The Great Plains of America, once the bountiful bread basket of the world, are a vast dust bowl. And those who cannot find food or shelter from the unrelenting sun are growing sick. Daily, it seems, we face new famines, new droughts, new flooding, as the Earth's climate inexorably warms. Daily, experts try to play God, desperate to determine what each new ecosystem will be, before it, too, is lost. This is the nightmare world of 2040 on this sad excuse for the planet we once called home.

The apocalyptic vision of science fiction? No. It is a world that exists, not in the imagination but in computer data used to formulate climate models — scale models of Earth, its winds and jet streams, high- and low-pressure systems, the data used by national weather forecasters daily. Some researchers and scientists have inserted into these model worlds the variables that humans have introduced, such as an additional 300 parts per million of carbon dioxide. Spin these models through 24 hours and you can reasonably predict the likelihood of rain in specific areas. Spin them through 438 000 hours, or 50 years, and you enter the world of Despair.

Scientists are by nature cautious, speaking of trends and indications and evidence within measurable limits that suggest what is likely or possible. However, confronted with the catastrophic implications of their data, some have broken this tacit vow of caution to take us on an unorthodox journey into the future, convinced that this is the nightmare world our children will inherit if we allow the greenhouse warming trend to continue.

WASHINGTON, D.C. — A.D. 2040 Fifty-two years ago, in 1988, when he was head of the NASA Goddard Space Institute, James Hansen made headlines when he warned the world that we had already embarked on an era of global warming. Now, he is living the reality of his computer projections:

> Right now, in 2040, the temperatures are about five degrees Celsius (9°F) warmer than they were in 1990. At the highest latitudes, the temperatures are almost 10°C (18°F) degrees warmer. You just have to look at a city like Washington, D.C., to see the impact. Washington is in the middle of a searing heat wave. The temperatures have been over 32°C (90°F) every day for the past three months. Fifty years ago, we saw that kind of weather maybe 30 days a year.
>
> The heat has had a dramatic impact on the amount of food available for the world's popu-

lation, because the frequency of droughts has increased enormously; as was predicted back in the 1980s, the warming up of the Earth's surface drove away rain-bearing clouds. Now, in 2040, we see the results. The great wheat-growing areas of the world, the bread baskets — the United States, Canada, Australia — have dried out.

For more than five decades, year by year, researchers like Jodi Jacobson at the Worldwatch Institute have documented the environmental destruction of the planet and the toll in human lives it was taking. But as Jacobson herself now admits, in this 2040 world, one in which as many as 60 million people are homeless, the warnings of impending chaos were sounded in vain:

One of the greatest causes of environmental refugees in our time and in the past 10 years, say since 2030 or so, has been the displacement of people as a result of rises in sea level. In Bangladesh, for example, we have had a sea-level rise of about 1.5 meters (5 ft.). More than 20 million people are now homeless there, and much of the land in that country is under water. There is daily fighting as Bangladeshis push their way into India, searching for food and a safe haven from the rising waters.

In the United States, large stretches of the Florida southern coast, including Miami and other areas that were rapidly being settled in the 1980s and 1990s, have now become dangerous places to live because rises in sea level have increased the incidence of storm surge. Now, we're getting killer hurricanes with winds in excess of 362 kilometers per hour (225 mph) and we're getting them more frequently than ever.

In Canada, seawater has tainted the supplies of fresh water in the Maritimes region, and people have been forced to flee inland.

Entire portions of Long Island that once were beach communities have disappeared. I myself grew up on a part of Long Island that no longer exists.

In the 1980s and 1990s, when I was quite young, the whole issue of global warming was controversial. There was a great deal of debate between policy makers and scientists on how far we should go, given the relative scarcity of data, and the uncertainties involved with global warming and sea-level rise, and the whole issue of how atmospheric gases were changing the Earth's temperature and climate. I realize now, sitting here in 2040 and looking back, that we were foolish not to act. We continued on the path that we had taken, of burning fossil fuels. We didn't move quickly enough to renewable energy sources or to energy efficiency. I think if we had known then what we know now — which, of course, is never the case — we would have moved more quickly to try to prevent global warming from occurring at the rate it has.

As farfetched as it may seem to us in 1990, the world described by James Hansen and Jodi Jacobson is on its way to becoming a reality, perhaps even sooner than 2040. Climate models show that with a doubling of current amounts of greenhouse gases such as carbon dioxide and methane (a National Academy of Sciences study in 1987 puts the doubling date at 2030), we could be looking at global warming of as much as five degrees Celsius (9°F) over the next half century.

We are already on our way to the world of 2040, says British climatologist Mick Kelly. The temperature of the planet is higher than it's been since record keeping began, and "the rate of change is likely to accelerate in the 1990s because of the tremendous increase in pollution we started to put into the atmosphere during the 1950s and 1960s." The six hottest years (including 1989) of the past 110 years all occurred in the 1980s. Kelly cautions that the next few years are a critical period: "We're breaking records already as far as global temperature is concerned. Any further warming is going to push us beyond the limits of natural variability, and that's when we can expect to see the first substantial impacts."

For decades it was thought that the oceans would absorb whatever excess carbon dioxide pollution our societies pumped out. We had an infinite sink into which our wastes could go; but the bubble burst on that comforting theory in 1957. That year two researchers, Roger Revelle and Hans Suess, working at the Scripps Institute of Oceanography in San Diego, gave us the first strong signal that we live in a finite world. They found, contrary to the wisdom of the day, that the oceans absorb only 50 percent of the excess carbon dioxide produced by man.

Today we know that roughly half of the carbon industrially produced appears to persist in the atmosphere, and we're adding to that load at an alarming rate. Each ton* of carbon discharged into the air results in 3.7 tons of carbon dioxide. There has been a worldwide rise in carbon dioxide emission of 10 percent since 1983. The "Big Warming" we expect is the result of human activity. We release heat-trapping molecules of such gases as carbon dioxide and methane and chlorofluorocarbons (CFCs) as the byproducts of our civilization; carbon dioxide from burning coal and oil, driving our cars, and

---

*Because the difference in weight between a tonne and a ton is not great, we have chosen to use *ton* in this book.

heating our homes, and from the destruction of forests; methane from cattle herds, and from rice paddies. We have a direct involvement in those emissions.

The Western world's desire for beef has doubled the cattle population in the past 40 years. There is now one cow for every four humans on the planet. Bacteria that break down the cellulose in the guts of cattle convert between 3 and 10 percent of the food the cattle eat into methane, which comes out the other end. It is estimated that the *flatulence factor* adds almost 100 million tons of methane to the atmosphere each year, conceivably enough to warm up the planet.

Rice paddies are implicated as the most important individual source for atmospheric methane. The roots of the rice plants seem to capture methane from the muddy bottoms, moving the gas straight through the plant into the air. As much as 150 million tons a year enter the air this way, from the 1.5 million square kilometers (579 195 sq. mi.) of paddies in the world.

Methane also is emitted by termites (about 5 million tons a year) and from bogs and marshes, burning forests and grasslands, putrefying waste sites.

All these sources together give off about 500 million tons a year. Most of the methane is destroyed by chemical reactions in the atmosphere, but each year 50 million tons more methane enter the atmosphere than leave it. Scientists suggest that by 2040 methane could be the prime greenhouse gas. We also add heat-trapping gases like CFCs from the consumer products we have made part of our lives: spray cans, air conditioners, and refrigerators. Like a pane of glass in a greenhouse, all these molecules let lots of sunlight in but prevent a large amount of heat from escaping the Earth's atmosphere into outer space.

The greenhouse effect is *us*, and it is specifically *us* in the Western world. More than 70 percent of the total global output of greenhouse gases comes from the industrialized world. Canada and the United States are two of the most energy-intensive countries in the world. Our per capita emissions of carbon are more than 16 times that of the developing world, and twice that of Japan and Europe.

With global warming will come a rise in sea level of as much as 1.5 meters (5 ft.). Most of the rise in water level will be the simple physical property of water — as the oceans warm, the water expands. Warming at the North Pole will free water now trapped in glaciers and icecaps, raising sea level even more. And if the West Antarctic ice sheet starts to slide into the ocean, the world could be looking at a 4.9 meter (16 ft.) rise in sea level. For years glacial mapping and radar surveys have suggested that this ice, a slab the size of India, is dangerously unstable. The fear is that global warming would weaken it further and set large chunks of it moving.

"We're facing a problem of survival that represents a threat to the planet as great as, or perhaps greater than, the threat of nuclear war," cautions Jim McNeil, secretary general of the 1987 U.N. Brundtland Commission on the Environment.

Something in the order of a third of the world's population and more than a third of the world's economic infrastructure are concentrated in coastal regions with altitudes below 1.5 meters (5 ft.). "All that is at risk over the next 40 to 60 years," says McNeil. Whole nations are at risk. A country like the Maldives, for example, and many other island countries are no more than 1.5 to 2 meters (5 to 6.5 ft.) above sea level. Today more than 75 percent of Americans, more than 180 million people, live within 80 kilometers (50 mi.) of a coast. Coastal land is some of the most fertile land in the world. Many of the rice paddies in countries such as China, Bangladesh, and Thailand exist right now at water level. Just a few feet of sea-level rise will threaten production of rice and other staples in these countries. If the rise is

1.5 meters (5 ft.), "we're looking at land loss greater than we've seen at any time in human history," says Stephen Leatherman, director of the Laboratory of Coastal Research at the University of Maryland. "And in fact it may change the configuration of some land. There will be so much total immersion that maps will have to be redrawn along many of the low-lying coastal areas, such as along the U.S. Atlantic and Gulf coasts."

The evidence is mounting that the nightmare world described here could become reality within our children's lifetime. On June 23, 1988, in the middle of a sweltering Washington heat wave, James Hansen made waves himself when he told a U.S. Senate committee that we are right now in the midst of the greenhouse effect; since the Industrial Revolution, first carbon dioxide, then these other molecules have been accumulating to form an invisible heat-trapping wall around the Earth, an atmosphere barrier that threatens all life on the planet.

On a day when the United States was experiencing what life will be like in the man-made hotbox, James Hansen testified before the committee that the Earth had become warmer in the past 20 years than during any other time on record. Even as he spoke, 1988 was on its way to becoming the warmest year ever recorded.

Hansen went on to say that his studies indicated that the average temperature of the Earth had increased by 0.6° to 0.7°C (1.08° to 1.26°F) since the Industrial Revolution. In short, what Hansen was saying was that by 1988 we had already embarked on the road toward the greenhouse world of 2040.

"It is time to stop waffling so much," Hansen pointedly told the assembled senators, "and say that the greenhouse effect is here."

James Hansen may have been the first scientist to put the "greenhouse effect" permanently on the front pages of newspapers and on everyone's lips, but he wasn't the first to describe the phenomenon. Almost 100 years ago, Swedish chemist Svante Arrhenius calculated that the coal burned since the Industrial Revolution might be releasing enough carbon dioxide to heat the planet. He predicted that a doubling of carbon dioxide in the atmosphere would lead to a rise in the Earth's surface temperature of four to six degrees Celsius (7.2°F to 10.8°F).

Arrhenius' conclusions went largely unnoticed for more than six decades, until 1958, when scientists at the Scripps Institute of Oceanography decided to follow up on his prediction. Charles Keeling, a graduate student, set up a carbon-dioxide-monitoring station at Mauna Loa in Hawaii. What he has been accumulating over the past 32 years is indisputable evidence that our oil- and coal-burning civilization is rapidly increasing the amount of heat-trapping carbon dioxide in the atmosphere — by more than 25 percent since the Industrial Revolution.

The first direct measurement of carbon dioxide levels showed them to be 300 parts per million (ppm) early in the twentieth century. When Keeling started his tracking in 1958, he measured annual concentrations of 316 ppm; by 1988, levels had climbed to 352 ppm, and by 1989, up two parts, to 354. Once that concentration of carbon dioxide passes 400 ppm, which it is predicted to do in the next 50 years or less, the atmosphere may be entering a rapidly changing state it hasn't been in before. And that is what has scientists worried.

Climatologist Stephen Schneider has been studying the global-warming problem for the past decade. Schneider points out that the concept of the greenhouse effect is not a scientifically controversial proposition; it is, after all, the process that gave life to the Earth originally. Schneider likes to talk about the Goldilocks theory: we just have to look at our sister planets Mars and Venus to see how important the global greenhouse is. Venus is too hot for life; Mars is too cold; the Earth is just right. Until now, atmospheric greenhouse gases like carbon dioxide have trapped just enough sunlight to keep the average surface temperature of the planet at a benign 15°C (59°F). Humanity is now tinkering with that life-sustaining balance. Schneider says:

> We already know enough to know that heat-trapping properties of gases are there, that gases are increasing and that we're going to heat the planet. The other thing we know — and, I think, virtually for sure — is that there is much better than an even betting-odds chance that by 2040 the world will be in a climatic regime unprecedented in the era of civilization. That means more than, say, two degrees Celsius (3.6°F) warmer than it was during the average of the past several thousand years. That rate of change is what has most of us concerned.

Nature changes the climate a degree or so over a millennium from time to time; five degrees over 5000 to 10 000 years. The last Ice Age was about five degrees colder than the present, so in the past 5000 to 10 000 years, we've warmed up about five degrees Celsius (9°F) on average — more like 10 or 20°C (18 to 36°F) in Canada. That temperature rise melted the mile-high ice sheet.

Says Schneider,

> The point is that the forests moved literally thousands of kilometers in response to this warming. The sea levels rose 100 meters (328 ft.). Species went extinct; others evolved. Warming revamped the ecological face of the planet, but it took 10 000 years to do it with *five* degrees.
>
> Now what are we talking about? The low end of what humans are talking about is one degree of warming over the next century. The high end is five to

10°C (9 to 18°F). So we're looking at changes that are *degrees a century*, whereas the natural rate of change is *degrees, at most, a millennium.*

What Stephen Schneider is telling us is that we're looking at rates of change at least 10 times faster than nature's; in 2040, it is quite possible that those rates will have escalated to 100 times nature's.

It is not only climatologists who are worried. Harvard biologist Edward O. Wilson, who has spent a lifetime documenting the life in the rain forests, says: "If we could say the world is going to get too warm to be comfortably sustainable, that species are going to go extinct in 10 000 years, then I don't think I would worry. But we're talking about *now* — not 10 000 years or even 1000 years, but *decades.*"

And we're talking about a change that will affect all life on Earth. "It would be sheer arrogance for any scientist to tell you that he or she understands the distribution of effects from that accelerated climate change," says Stephen Schneider.

How will diseases change? How will ecosystems be affected? What will happen to crops, water resources, forest fires? The more rapidly the climate changes, the less likely it will be that we can predict; and to adapt, we must be able to determine in time what is and will be happening. Nature needs even more lead-time to adapt than we do. Quite simply, that means things will go extinct in place unless we start becoming ecological engineers and transplanting them, with who knows what consequences.

A.D. *2040*   Joel B. Smith of the U.S. Environmental Protection Agency knows how prophetic Stephen Schneider's words have proved to be. In 2040, scientists do not have the luxury of indulging in long-range planning and experimenting. Smith no longer trusts the reasoned calm of scientific exposition. He speaks now with passion about the changes wrought over the past 50 years:

Fifty-one years ago, in 1989, I wrote a report to Congress on the potential effects of global climate change on the United States. Sadly, I am now living through the horrible projections we made. We're seeing the decimation of forests across the globe. We've been tracking the problem for the past half century or so as temperatures have risen.

What we're seeing is that some of the forests are migrating northward toward the poles, but at a much slower rate than they're dying back on the southern border; they can't move fast enough to keep up with the climate change. We were warned about that very thing more than 50 years ago by Michael Soule, then at the University of Michigan. Soule would get a big laugh from students when he graphically demonstrated what would happen to trees as the climate warmed. His example was a potted plant that he'd kick across the floor to show how quickly it had to move to survive. A tree would literally have had to get up and walk away. Based on rates at which forests migrated after the last Ice Age, Soule figured out that with just a two-degree change in temperature the boundaries of a species of trees would have to move at rates of the order of 1 meter (3 ft.) an hour in order to stay in an area with a livable climate. Of course that has proved to be impossible, and we're seeing the forests around the world disappearing before our eyes. This is most evident in northern areas of Canada, Scandinavia, and the Soviet Union, where the warming has been the greatest, and there's been the greatest reduction of boreal forest because the trees just don't have anywhere to go. They can't migrate over the icecaps.

We're desperately trying to replant forests everywhere, but it's difficult, given the scale of this problem. It's hard to keep up. Scientists and planners and resource managers are worried because we're told the rate of change is only going to accelerate.

What do we plant for tomorrow? What do I plant today if I've got a tree that's got to last 50 or 100 years and I don't know how much warmer it's going to get, because that heat just doesn't stop?

Fisheries have been significantly affected. Higher temperatures have caused the ocean fisheries to migrate northward, and people are trying to make adjustments, either traveling farther in fishing boats just to catch the fish they were used to catching, or trying to change their dietary habits to keep up with the rate of change.

In some ways, inland fisheries pose an even greater problem: those fish can't migrate on their own from one lake to another. People have been trying to respond by restocking lakes with fish, moving into northern areas. But of course they're running into problems regarding what to put where. In that sense they're almost trying to play God, trying to figure what the new ecosystems are going to be like.

The biggest problem we're facing right now, in 2040, is that the climate has not stopped warming; it's not a matter of our catching up and adapting. Because people did not take action in the past to slow down emissions of greenhouse gases and because we're continuing to burn fossil fuels, cut down whatever forests are remaining, and use chlorofluorocarbons, the rate of warming just keeps getting faster and faster. It's now becoming very difficult for man — one of the most adaptable, flexible creatures on the planet — to keep up with the rate of warming that we're seeing, and it's only going to become harder.

That is the case in Canada in this year almost halfway through the twenty-first century. As John Last, a physician who specializes in public health, looks around Canada in 2040, he sees a country where people are picking their way through the ruins of what was once a civilization:

Canada was less damaged by the disastrous crop failures caused by the greenhouse effect than were other countries in the grain-producing parts of the world. Canada's climate actually warmed up a little and got slightly wetter, whereas in the United States, for example, it dried up.

By the time large numbers of ecological refugees began moving into Canada from the United States, from the drought-stricken Midwest, we'd already absorbed a good many million refugees from other starving parts of the world. By 2020, Canada's population had doubled from what it had been in the late 1990s, and was somewhere between 60 million and 70 million. Nobody knew for sure because so many people were rootless. The three largest cities — Toronto, Vancouver, and Montreal —

had populations of well over 10 million each, the majority living in shanty-town slums on the outskirts or in the decayed inner-city cores. But in addition there were refugee tent cities in various parts of the country, and an estimated two million or three million people were roaming the country, searching for food and shelter.

Food shortages had been common for a long time, since the early 1990s in fact, but by the early twenty-first century, those food shortages had become quite serious famines. Famine leads to reduction of resistance to infections, so epidemics have come with the famines.

As well as the epidemics, we are experiencing other kinds of problems unprecedented in the recorded medical history of North America. The changed climate led to a new distribution of insect vectors of disease: by the year 2025, malaria and viral infections, such as yellow fever, were endemic in central Canada and throughout most of the Mississippi Valley and the eastern seaboard of the United States.

Unfortunately the health-care system in the late twentieth century had become a technology-oriented service. Very few physicians had any interest in promoting health and preventing disease, and these aspects of the physician's work played a very small part in medical education. By the end of the twentieth century, however, it had become apparent that our high-tech kind of medical care was irrelevant to the human situation.

It took the medical and other health-related professions almost a generation to recognize the need to reassess priorities, and to place maximum emphasis on educating physicians and other health workers to deal with public-health problems of the kind that had existed in developing countries in the early twentieth century, but were unimaginable in the industrialized world as late as the early 1990s. As we've seen, they've become increasingly imaginable from that time on.

As I look around this world of 2040 and look back to my youth in 1990, I am filled with sadness that the messages some of us tried to convey about the urgent need to change our ways went unheeded. All around me are a great many people suffering not only from the effects of food shortages that arise from the greenhouse effect, but also from overexposure to ultraviolet radiation. In the past 75 years, the amount of ultraviolet radiation penetrating to the Earth's surface has progressively increased. As long ago as the mid-1970s, the National Academy of Sciences in the United States, after reviewing the evidence, reported that there were grounds for concern about the deterioration of the ozone layer.

Our man-made chlorofluorocarbons from our refrigerators, our air conditioners, our spray cans, and the polystyrene containers we used back then were systematically destroying the fragile shield of ozone around the Earth that had acted as a buffer against the sun's ultraviolet radiation.

Serious biological effects were apparent by the early 1990s. The consequences for humans were increased incidence of skin cancer, increased frequency of severe damage to the eye — cataracts and damage to the cornea — and damage to the immune system, which made people more susceptible to infections of all kinds and also to certain varieties of cancer.

As a physician, I feel myself to be helpless in dealing with these horrible problems. Now, unhappily, I look at a world where my grandchildren are growing up and seem unlikely to be able to cope with the disasters that they have inherited from us.

Ironically, the ozone catastrophe of 2040 had its roots in one of the scientific marvels of the late twentieth century. As is true of so many other technological creations of human ingenuity, the visible benefits of CFCs cloaked invisible threats. But the problems were not to appear until much later. None of that was known on a day in April 1930 when chemists were caught up in the excitement of a discovery made by one of their peers:

Thomas Midgley smiled to himself as he looked around the room. It wasn't a bad turnout for his session at the American Chemical Society meeting. Obviously even chemists like a show. Dr. Midgley placed his face over the beaker containing his latest chemical triumph and inhaled deeply. As he lifted his face, he could hear the murmurs grow, but when he turned and extinguished a candle in a second beaker by exhaling the gas gently through a tube, the room erupted in laughter and applause. Tom Midgley had done it again. He'd found a safe, non-toxic, non-inflammable, cooling gas that would revolutionize refrigeration and air conditioning. Freon had made its public debut.

Forty years later, Thomas Midgley's concoction and a family of similar compounds called chlorofluorocarbons were being produced at an incredible rate. In 1976, U.S. manufacturers were producing 340 million kilograms (750 million lb.) of CFCs a year — 55 percent used as propellants in aerosol sprays.

By this time, CFCs were being used for everything from solvent for cleaners of silicon chips for computers, to air conditioning in cars, to blowing agents in the manufacture of polystyrene cups, egg cartons, and fast-food containers. They were amazingly useful, cheap to manufacture, non-toxic, non-inflammable, and chemically stable. They didn't seem to react with anything. They were the perfect man-made compound.

In 1972, at another scientific meeting, the topic of fluorocarbons came up again, this time during an informal discussion about their incredible chemical stability. F. Sherwood Rowland of the University of California kibitzed in the hallway with his colleagues. Perhaps that quality could be used to trace large-scale wind patterns, someone mentioned. At this point it was merely a scientific curiosity that the quantity of fluorocarbons in the atmosphere was almost equal to that produced by the chemical industry. They would, of course, eventually be broken down by ultraviolet light, observed Rowland. Or would they? This casual conversation about a basic scientific point led Sherry Rowland and his colleague Mario Molina to try to unravel the fate of fluorocarbon molecules. They had no premonition of the environmental importance of the question they were seeking to answer.

In December 1973, when Rowland and Molina first perfected the calculations that showed the destructive power of the fluorocarbons, they couldn't believe what they'd uncovered. While the fluorocarbon molecules do, indeed, remain stable in the lower atmosphere, once they get high enough, 50 to 80 kilometers (30 to 50 mi.) above the Earth, the ultraviolet light from the sun breaks them apart and knocks off chlorine atoms. Those chlorine atoms that are released then launch an attack on the ozone layer: each chlorine atom can destroy up to 100 000 molecules of ozone.

Rowland remembers being stunned by these results: "Our immediate reaction was 'We have made some huge error. This is much too big a number.' We hadn't heard anything before about this possible attack on ozone; no one was talking about it. Your first reaction is that if a number is that big, somebody would have said something."

When Rowland and Molina announced their results, they shocked their scientific peers by calling for an immediate ban of CFCs in spray cans. They stated that CFCs were weakening the ozone layer enough to cause a marked increase in skin cancers and could contribute to global climate change. But they had lined themselves up against a $28 billion-a-year industry.

For years Rowland and Molina carried their story to anyone who would listen. The powerful CFC industry discounted their hypothesis and tried to discredit them. In August 1977, the president of one aerosol-manufacturing firm suggested that criticism of CFCs was being "orchestrated by the Ministry of Disinformation of the KGB."

A year or so later, computer simulations pretty much confirmed Rowland and Molina's theory that CFCs could be chemically broken down in the upper atmosphere by the sun's ultraviolet radiation, and their hypothesis on ozone depletion was accepted. When, under pressure from a public now boycotting aerosols containing CFCs, governments, including those of Canada and the United States, began to regulate the use of CFCs in the manufacture of spray cans, it seemed to many that the problem had been pretty well solved. It hadn't been.

CFC use grew in other areas — as coolants in commercial and car air conditioners and refrigerators; foaming agents in foam insulation, packaging, and cushioning; solvents that clean solder from electronic components — while CFC-like compounds called halons, which are 10 times more potent as ozone depleters, were developed and used in fire extinguishers. All the while, scientists carried on scholarly arguments over just how much of the ozone layer could be depleted, and penciled in estimates ranging from 2 percent to 20 percent.

If CFCs were going to affect the ozone layer, the consensus was that there would be a gradual depletion. No one was prepared for the announcement in 1985, by a team of British scientists, of an enormous hole punched out of the ozone layer over Halley's Bay on the coast of Antarctica.

The recorded dips in the ozone concentration were so large (up to 40 percent) and the idea of a continent-size hole so unexpected that several research groups, including the British Antarctic Survey, NASA, and the U.S. National Oceanic and Atmospheric Administration had not wanted to believe their own findings. Each group had waited for independent corroboration and, hearing no confirming reports, had sat on the results for years, attributing readings of enormous seasonal loss of Antarctic ozone to instrument or computer failure.

Contrary to the expectations of scientists, the layer protecting the Earth wasn't just thinning gradually. The CFC-ozone reaction wasn't linear: that is, it wasn't simply a matter of a little chlorine thinning the ozone layer a little bit, a little more chlorine thinning it a little bit more. Instead, an immense hole was suddenly punched in that section of the Earth's atmosphere. And even more ominous was the fact that the hole had occurred over the remotest part of the globe, far from where the CFCs had been liberated. What the ozone hole underscored was our fundamental ignorance of how our actions affect atmospheric chemistry.

That was 1985. Over the next four years, concern for the ozone layer prompted an unprecedented and quick global response. International meetings were held first in Vienna, then in Montreal in 1987 where an international protocol was drawn up that called for a 50 percent cut in CFCs by the year 2000. By 1989 it was clear these steps were inadequate. A Helsinki meeting in 1989 discussed a total ban of chlorofluorocarbons by the end of the decade. In June 1990, the 56 nations that had signed the

Montreal protocol met in London. Facing new research that showed the ozone layer was being depleted much faster than previously thought, the 56 agreed to the total ban by 2000. After the United States, Japan, and the Soviet Union balked at setting an earlier timetable, Canada and 13 other countries issued a separate declaration, saying they intend to complete the phaseout of CFCs and halons by 1997.

But even the measure of apparent global commitment to ozone preservation is too little, too late. As Sherry Rowland admits, "Even if we had a total ban tomorrow, and there were no further emissions, we would expect things to get worse until about the end of the century and then gradually recover over two centuries." And every year of continued global CFC use increases the projected recovery time for the ozone layer by $3\frac{1}{2}$ years.

Most CFCs remain in the atmosphere for 75 to 100 years; therefore, virtually all the CFCs that have ever been released — roughly 800 000 kilograms (1 763 668 lb.) to one million tons of CFCs were added each year during the early to mid-1980s — are still in the atmosphere.

And even the proposed replacements for CFCs — HCFCs and HFC-134A — are going to cause problems. Although much less dangerous per molecule, HCFCs contain some chlorine, which can damage the ozone layer; and HFC-134A is a heat-trapping gas that can add to the greenhouse effect.

Meanwhile, the Antarctic ozone hole has drifted over Australia, while another has been found over the Arctic. In the winter of 1989, the Antarctic ozone hole was the largest it's been since it was discovered. The hole covered an area two and a half times the size of Canada. Within that area 45 percent of the ozone was gone. In June 1990, Australian scientists announced that yet another hole had been discovered over MacQuarrie Island in Australia. The damage was spreading. Even outside the "holes," the layer has been thinning. A report released at the London meeting showed a 10 percent reduction in ozone since 1967 over the middle latitudes of Europe and North America. The observations were made over Switzerland, Germany, and Canada.

A U.N. Environment Program draft report estimates that just a 1 percent decrease in ozone will lead to a 3 percent increase in the rate of skin cancer worldwide and will result in 100 000 additional cases of blindness caused by cataracts.

This report also predicts that a depletion in ozone will increase the outbreaks of infectious disease, because normal immune systems become less effective when they have been exposed to increased ultraviolet "B" radiation. That means more widespread and frequent outbreaks of measles, herpes, tuberculosis, and leprosy can be expected to occur. Compounding this problem, the effectiveness of vaccines is compromised when they are injected into skin that has been exposed to ultraviolet "B."

The U.N. Environment Program report points to other serious problems caused by increased UV light because of the thinning ozone layer: food production on both land and sea will decrease. On land, many plant species will not grow, flower, or germinate in increased ultraviolet light. In water, phytoplankton, microscopic plants that live in lakes and oceans, are essential to climate control and the stability of the food chain. These plants are very sensitive to increased levels of ultraviolet light. Although UV does not penetrate water, surface phytoplankton kills can be considerable. These microscopic organisms are the base of the marine food chain. If they go, the food pyramid could topple. Phytoplankton also absorb as much as half the carbon dioxide produced globally each year. If they begin to die, global warming will escalate.

Now, take all the data we've got on ozone depletion and insert them into the computer models for future climate change and this is what we get:

BRISBANE, AUSTRALIA — *A.D.* **2040** For Ian Lowe, the director of the Science Policy Research Centre at Griffith University in Brisbane, Australia, the world of 2040 is one of devastating atmospheric depletion:

> Australia was one of its first victims, but there was little we could do about the ozone hole. By the middle of the 1990s it had spread rapidly from Antarctica to extend over parts of Australia and New Zealand. In 2003, we had an epidemic of malignant melanoma, a virulent and often fatal skin cancer, triggered by the depletion of ozone.
>
> The ozone hole over Australia changed the national character of the country. Up until the 1990s we had always been an outdoors people; our culture was really our sport. But with the dangerous levels of ultraviolet light that we started to record in the first few decades of this century, we gradually moved inside. If you look around Australia today, you hardly see people outdoors unless they have to be. We spend our lives inside.
>
> In the 1990s Australian governments started to take tough measures to deal with the rising global temperatures and sea-level changes. It's difficult now to remember what life was like in the 1980s. People actually used to fuel cars or outboard motor boats and drive them around mindlessly on Sunday afternoons, just for something to do.

By the late 1990s the first Green party government of Australia brought in a carbon tax, and almost overnight the use of fuel plummeted dramatically. People's behavior radically changed as the price of gasoline doubled, then tripled, then quadrupled. Tax revenue was used to fund free public transit and to build an extensive system of bicycle paths.

The big crunch came when private cars were banned from central urban areas. That happened in 2005, but by then, Australians were ready to bite the bullet because they were already seeing the drastic effects of the greenhouse effect. We started to wage an all-out war against carbon dioxide emissions to try to slow down the warming trend. Free solar domestic hot-water systems were supplied to all houses on the mainland. These measures resulted in a massive drop in the use of fossil fuels.

When I tell my grandchildren the things that people used to do in the twentieth century, they're speechless with disbelief. They can't believe that people bought food wrapped in cardboard and plastic packaging that was thrown out. They can't imagine that people dumped sewage in the sea, and knowingly polluted the air and water and soil. They can't believe we sacrificed their world by not acting internationally on global warming soon enough. The change in consciousness and environmental awareness has been dramatic and irreversible. Tragically, we see now, in 2040, that it came too late for much of the world.

It is said that hindsight is 20/20 vision. It's easy to look back from the year 2040 and see what we should have done. But the computer models of that future world are not enough to move governments to action in 1990. There is a substantial gap between green rhetoric and action by all our Western governments. This is especially true when the problems have not yet been clearly defined, yet pose enormous risks. Take the greenhouse effect.

In October 1989, a strange and protracted argument took place in White House offices. In one camp was the Environmental Protection Agency head, William Reilly, and standing off against him was the powerful duo of President Bush's science adviser, Allen Bromley, and the White House Chief of Staff John Sununu. Over the next several weeks, hints of what happened leaked out in press reports.

What emerged was a picture of a White House trying desperately to minimize all the warnings about the greenhouse effect. It was the official opening volley in the *greenhouse denial syndrome*. The United States was scheduled to participate in an international conference in the Netherlands that November. The purpose of the meeting was to stabilize carbon emissions at 1988 levels by the year 2000, and to study the feasibility of a 20 percent cut worldwide by 2005.

Bromley and Sununu and others were worried that the United States, which produces one-fifth of the world's carbon dioxide emissions from its cars, industries, and heating, could suffer economic damages if stabilization levels or reductions were agreed to. What was at stake was the nation's major supply of energy — coal and oil, the two fossil fuels that contribute the greatest amount of carbon dioxide to the atmosphere.

The argument that October day was over whether the United States should even send a representative to the Netherlands meeting. Reilly not only wanted to go but wanted to go with a presentation that would put the United States in some kind of leadership position in dealing with global warming. In the end, Reilly achieved a hollow victory: the United States attended, but allied itself with Japan and the Soviet Union to scuttle any international attempt to set goals for curtailing carbon dioxide emissions.

The White House was grasping at straws to downplay global warming, and therefore had bought, lock, stock, and barrel, the conclusions of a report that "essentially wishes away greenhouse warming." That document was prepared by a trio of scientists — William Nierenberg, Robert Jastrow, and Frederick Seitz — who a few years earlier had been staunchly defending SDI, or Star Wars.

None of the three are prominent greenhouse-effect researchers, yet they declared it null and void in their report for the Washington-based George C. Marshall Institute. The warming recorded over the past century was due to the sun's variations and not carbon dioxide buildup, they speculated, and they went so far as to claim that we can actually expect a mini-ice age in the next century.

Greenhouse scientists were dumbfounded. "That 0.5 to 0.6 degree warming we've had until now is not the issue," contend climatologist Stephen Schneider and others. What is alarming about greenhouse warming is not the increase in the past century but what is happening right now. Greenhouse-gas emissions — carbon dioxide, methane, and chlorofluorocarbons — are increasing so rapidly that it is "dead certain" when enough are released we will be into an era of unmanageable global change.

Those gases trap heat, and they aren't going anywhere but up. Pure and simple logic dictates that when enough of them build up, global warming is inevitable. As Schneider said of the 1989 report: "I have heard that Sununu is holding the Marshall Institute report up like a cross to a vampire, fending off advocates of slowing down global warming."

Stories like this one are familiar to Schneider. He and his colleagues are constantly being challenged by policy makers as to how realistic the greenhouse predictions are. Schneider likes to tell this story when the predictions are challenged:

A lot of people say, "Aren't there mechanisms in the system that could make you guys wrong? I mean, you're predicting three to five degrees' warming in 50 years. How do you know it isn't going to be one degree?"

Indeed it might. In fact, I was talking to the U.S. State Department undersecretary last year about the need for international taxes on fuels and so forth, and he said, "Wait a minute, that's a pretty serious step. You guys could be wrong. Couldn't there be a buffer? Couldn't more water evaporate, make more clouds? More clouds are brighter and that reflects away sunlight, so-called negative feedback."

And I said, "You're right!" And he said, "Gee, you're pretty easy. I thought you were going to be a tough guy to debate with." I said, "Well, now please tell me what is the probability that this buffer's going to be out there?" He was a little unsure about that. I said, "You think it's one chance in 10 or how about one in three?" He said, "Okay, I'll take one in three." I said, "I'll do better than that. I'll give you one in two. I'll give you a 50 percent chance that we've overestimated it by a factor of two or three — that's about the most we could." He said, "Okay." I said, "Good, then you've got to give me a 50 percent chance that either we're right or we've underestimated by a factor of two or three."

Because clouds don't just get wider, which cools the planet, they can get taller, which warms the planet. Or, if you warm the planet, there's organic matter in the soil, which, through microbial action, decomposes more rapidly when it gets warmer, releasing carbon dioxide faster; and then there's all this methane sitting under the tundra and off the continental shelves that can get dumped in. Those are positive feedbacks that would exacerbate the warming.

So we've got feedbacks, both positive and negative, unknown, and the bottom line is that we're insulting the environment faster than we know what we're doing.

So I said to the official, "If you don't want to act, what you're really doing is taking a coin and writing one face on it that says 'unprecedented climate change in the middle of the twenty-first century' and flipping it, because that's the gamble that we're facing."

Politicians and bureaucrats are hoping against hope that the climate modelers are wrong, and that the human-wrought changes to our atmosphere won't amount to anything so that they won't have to act on carbon dioxide emissions. They're hoping that there are natural feedbacks, natural ways that the Earth can stabilize itself that we don't yet understand. Ralph Cicerone, chairman of the department of geosciences at the University of California at Irvine, calls this the "magic thermostat syndrome." The idea is that there is some magic regulating system that we haven't yet discovered that will absorb any amount of insult to and change in the environment and somehow keep the Earth's temperature roughly where it's been. Politicians also advance another argument: there's so much uncertainty in the greenhouse predictions that we shouldn't try to act on policies based on such uncertainties. So they opt to do nothing. And for their part, climate modelers are hoping against hope that the positive feedbacks they factor in to their global climate models won't kick in, because if they do, it will be too late.

A.D. **2040**   Fifty-two years ago, Jeremy Leggett gave up his position as a geology professor at the University of London to become the director of science at Greenpeace U.K. He did it because he realized that the Earth he taught his students about —

a planet that changed over agonizingly slow eons — no longer existed. Leggett now observed abrupt changes occurring in the planet that he so carefully studied, and he became an activist so he could alert the public. That was in 1989 . . . but now, in 2040, Jeremy Leggett sees the world around him a testament to denial:

When I look back at 1990, the thing that strikes me most is the unjustified optimism of some scientists and policy makers at that time. We managed to persuade ourselves that we were looking at temperature increases of between 1.5 and 5 degrees Celsius (2.7 to 9°F), and of course all that was predicated on not looking at worst-case scenarios.

There were things that now, with the benefit of hindsight, looked so obvious. Once the temperatures had risen a few degrees, all the methane, trillions of tons of methane locked up in the tundra and underneath the shallow seas in the Arctic, started to melt out of the methane clathrates (lattice-like geological structures) and float up into the atmosphere, becoming a greenhouse gas. As a heat retainer, methane is 20 to 30 times worse, molecule for molecule, than carbon dioxide.

And, of course, the greenhouse effect then started to run away with itself and produce temperatures the likes of which we didn't really expect, even in our worst dreams. Ecosystems can't move fast enough. Vegetation belts during the last glaciation, when temperatures were changing at rates of a couple of degrees Celsius (3.6°F) in 1000 years, instead of a couple of degrees Celsius a decade, had a hard enough time migrating from the low latitudes to the high latitudes. Now, of course, they can't keep pace with the temperature changes. There are too many direct stresses on ecosystems that are very finely balanced. We've seen that effect in devastating famines all over the world and in the collapse of many industrial societies that have got a strong component of agriculture built into their economies, and really we don't look as though we're going to make it very far into the twenty-first century.

Stabilizing the gaseous composition of the atmosphere is the most urgent problem we're facing in 1990, says George Woodwell who heads the Woods Hole Research Center on Cape Cod. "We've already made a commitment to warming the Earth by allowing heat-trapping gases to accumulate. The more the Earth warms, the more carbon dioxide and methane are released from the soil and from the destruction of forests, and the more the Earth will warm after that."

Right now we release from burning fossil fuels between five billion and six billion tons of carbon annually. We release another two to three billion from deforestation. Woodwell says his institute's calculations show that within 10 years "we'd have to reduce fossil-fuel use by 75 or 80 percent to get to the point where we are no longer building up carbon dioxide in the atmosphere from that source." That's a 75 to 80 percent reduction just to stall the warming trend — the world can't even get together on a 20 percent reduction. "The reduction has to be started right now," Woodwell says emphatically; "it cannot be started soon enough. It's a bit late for discussion and time for action."

The nightmare world of 2040 reaches back across time to warn us that everything we now take for granted will change, if not in our lifetime, then our children's, and in ways that will rob us of our dreams for tomorrow. Right now that threatened future is calling out to us to understand that these visions are not flights of science fiction or of fantasy. They are serious, sober-minded projections of how the world is going to change. You can find them in Stephen Schneider's book, *Global Warming*; you can find them in Ian Lowe's book on the future of climate; you can find them in Joel Smith's report to Congress for the Environmental Protection Agency; and you can find them in the Worldwatch Institute's annual reports on the state of the world that Jodi Jacobson helps to write. Today, though 2040 exists only in the computer models of climatologists, the future it offers is not an abstract concept but a scientific reality.

What if you had it in your power to change that future. Would you do it? There's a poignant moment in the movie *Superman*, when Lois Lane is killed and the grieving Superman flies above the Earth, faster and faster, reversing the Earth's rotation and reversing time to regain that moment when Lois Lane's death was preventable. It's a haunting image precisely because it reminds us of those times in our own lives when we've wished desperately that by sheer will we could turn back the clock, forestall a crisis, a loss, a tragedy.

"If only," we think at such times, "if only I could have known what was coming, then I could have prevented it." We have the opportunity to do that right now. We have just seen the world that we are leaving our children. We do have it in our power to change that future, to make sure the world of 2040 doesn't happen. We can do it by confronting the hard facts of what we're up against, the depth of the crisis, then making the tough decisions that will save us.

TWO

*H O W   D I D   W E
C O M E   T O   T H I S ?*

───────────────────────────────

There's a strange phenomenon that biologists refer to as "the boiled frog syndrome." Put a frog in a pot of water and increase the temperature of the water gradually from 20°C to 30°C to 40°C... to 90°C and the frog just sits there. But suddenly, at 100°C (212°F), something happens: the water boils and the frog dies.

Scientists studying environmental problems, particularly the greenhouse effect, see "the boiled frog syndrome" as a metaphor for the human situation: we have figuratively, and in some ways literally, been heating up the world around us without recognizing the danger.

Psychologist Robert Ornstein, co-author of *New World, New Mind*, points out that those people who have been sounding warnings receive the same response from us as would someone attempting to alert the frog to the danger of a rise in its water temperature from, say, 70° to 90°C (158° to 194°F). If the frog could talk, he would say, "There's no difference, really. It's slightly warmer in here, but I'm just as well off." If you then say to the frog, "If the heat keeps increasing at that rate, you will die," the frog will reply, "We have been increasing it for a long time, and I'm not dead. So what are you worried about?"

"Our situation is like the frog's," says Ornstein. Today, despite the fact that researchers using the most sophisticated atmospheric-monitoring equipment in the world are telling us that our future is at risk, we — as individuals and as governments — ignore or minimize the warnings.

The frog has a fatal flaw, explains Ornstein. Having no evolutionary experience with boiling water, he is unable to perceive it as dangerous. Throughout their biological evolution, frogs have lived in a medium that does not vary greatly in temperature, so they haven't needed to develop sophisticated thermal detectors in their skin. The frog in the pot is unaware of the threat and simply sits complacently until he boils.

Like the simmering frog, we face a future without precedent, and our senses are not attuned to warnings of imminent danger. The threats we face as the crisis builds — global warming, acid rain, the ozone hole and increasing ultraviolet radiation, chemical toxins such as pesticides, dioxins, and polychlorinated biphenyls (PCBs) in our food and water — are undetectable by the sensory system we have evolved. We do not feel the acidity of the rain, see the ultraviolet radiation projected through the ozone hole, taste the toxins in our food and water, or feel the heat of global warming except, as the frog does, as gradual and therefore endurable. Nothing in our evolutionary experience has prepared us for the limits of a finite world, one in which a five-degree climate change over a matter of decades will mean the end of life as we have known it on the planet.

How did we come to this? How did we plan our own obsolescence? The answer lies in millennia of human history, a surprisingly brief chapter in the chronicle of the planet. You can see just how brief if you use a standard calendar to mark the passage of time on Earth. The origin of the Earth, some 4.6 billion years ago, is placed at midnight January 1, and the present at midnight December 31. Each calendar day represents approximately 12 million years of actual history. Dinosaurs arrived on about December 10 and disappeared on Christmas Day. *Homo sapiens* made an appearance at 11:45 P.M. on December 31. The recorded history of human achievement, on which we base so much of our view of human entitlement, takes up only the last minute of that year.

The dinosaurs had a fortnight of supremacy on this planet before they were eradicated by some environmental catastrophe. We have had 15 minutes of fame. And in that short period we have transformed the world. In fact, *Homo sapiens* has managed to extinguish large parts of the living world in a matter of centuries.

For the past million years, our biological makeup has changed very little: we are essentially the same creatures as those that emerged along the Rift Valley in Africa. But culturally we are a completely different genus. Where once we lived as part of the natural world, we now seek to conscript it in the service of our ends. And the result has been that we have altered the ecosphere in which we evolved into a form that cannot ensure our survival.

Compared to that of other mammals, human physical prowess is not very impressive. We are not built for speed: a mature elephant can easily outrun the fastest sprinter in the world. We have no physical arsenal for self-defense — no claws, fangs, or horns. Our sensory acuity is limited: we can't see over great distances, as the eagle can, or at night like the owl; our hearing is not highly developed, as is the bat's; we can't sense ultraviolet light, like a bee. Indeed, our main survival strategy has been the development of a 1600 gram (3.5 lb.) brain, capable of complex thought.

Out of that brain evolved a mind possessed of self-awareness, curiosity, inventiveness, memory. It enabled us to recognize patterns and cycles in the natural world — rhythms of day and night, seasonal changes, tides, lunar phases, animal migrations, plant successions — and to make out of this predictability a mode of existence for ourselves in what we saw as our natural environment. For millions of years our ancestors shared the grassy savannahs with other mammals and birds and plant life. We learned to "read" the alarms sounded by species more attuned to danger than we were. Other living things became our "early warning system."

Secure in our ability to rely on the natural world to troubleshoot for us, we began to shift from passively surviving to actively seeking to improve our quality of life. We made crude tools to facilitate hunting, and even our crudest ones were sufficient to make our early forebears deadly predators. The arrival of paleolithic peoples in North America across the Bering land-bridge perhaps 100 000 years ago was followed by a north-south wave of extinction of species of large mammals.

Then, between 10 000 and 12 000 years ago, a major shift in human evolution took place — *Homo sapiens* became a farmer. We learned how to domesticate and cultivate plants and in doing so made the transition from nomadic hunter-gatherer to rooted agriculturalist. Once we knew how to ensure that we would have food, settlements grew up, and we started to exploit nature to serve our needs. We began to manage the planet, and the water in our pot heated up a few degrees.

Man the farmer was far more destructive than man the hunter. Early agriculture changed the landscape of the planet. While it fostered the rise of cities and civilizations, it also led to practices that denuded the land of its indigenous flora and fauna and depleted the soil of its nutrients and water-holding capacity. Great civilizations along the Indus and Nile rivers flourished and disappeared as once-fertile land was farmed into desert. We read the devastating signs as an indication that it was time to move on. We were few in number and the world was infinite.

The Industrial Revolution stoked the fire under our pot. The replacement of muscle power by machines, driven by wood and coal, then oil and gas, and finally nuclear fission, drowned out all warning signals under the roar and clatter of a culture heading inexorably toward the new Eden of growth and progress.

Once it took Haida Indians in the Queen Charlotte Islands more than a year to cut down a single giant cedar. When the Europeans arrived with the technological know-how — the two-man saw and steel ax — the task was shortened to a week. Today, one man with a portable chain saw can fell that tree in an hour. It is that explosive increase in technological muscle power that has enabled us to attack the natural world and to bludgeon it into submission.

The massive application of technology by industry has been accompanied by an unprecedented increase in human population. With the conquest of the major causes of early death — sepsis, infectious disease, and malnutrition — world population numbers have exploded.

We are now the most ubiquitous large mammal on Earth, and armed with technological might, we have assumed a position of dominance on the planet. We have gained the ability to destroy entire ecosystems almost overnight, with dams, fires, clearcut logging, agricultural and urban projects, and mining. And our self-awarded mandate to do so is predicated on the assumption that we possess the knowledge to manage the environment, that nature is sufficiently vast and self-renewing to absorb the shocks we subject it to, and that we have a fundamental entitlement to nature's bounty. That has been our history, but the reality is different. Our machinery, based on fossil-fuel consumption, produces carbon dioxide in quantities that exceed the ability of the natural world to absorb it, as we inject almost six billion extra tons of carbon annually into the upper atmosphere, just from the burning of fossil fuels. As well as increasing naturally occurring compounds — methane, nitrous oxide, sulfur — the chemicals industry has introduced man-made compounds such as CFCs, while the pulp and paper industry generates chloro-organics such as furans and dioxins. Technology enables us to harvest previously unattainable "resources" in quantities far in excess of their regenerative capacity, and we pay no attention because, through the span of human history, we have considered those resources infinite and have believed that the human mind would always provide solutions to our problems. That belief has entrenched in us a mindless acceptance of certain truths and an equally mindless lack of comprehension of the consequences.

But history also tells us that both our ignorance of how the world works and our destructiveness have cast us in a dangerous role. There is a revealing story about all this that sounds like a parable, but unfortunately is true:

Once upon a time, there came into being an exotic land. Cut off from the rest of the planet, it gave birth to a glorious but vulnerable world that was blanketed in softness. Deep-rooted, pliant grasses flourished in loose soil. If you had been there you could have plunged your hand inches into the rich earth. Everything that moved on the land moved gently, including strange animals that carried their young in pouches. Everything that trod the land had soft paws. And the land was benign and fruitful for all that lived there. Isolated through millions of years of time, the land was inviolate.

And so it was, for too many generations to count.

Then suddenly, almost in a moment, everything changed. Two hundred years ago the white man arrived on this strange alien continent and chose it for his own. The otherness of the landscape disturbed the newcomers, and they sought comfort in the creation of the familiar, transforming their surroundings to duplicate the memory of a place called home, a place called England. It was the beginning of a new era for this exotic world, one the Australian author and farmer Eric Rolls calls the time in which "they all ran wild."

Within six years of English settlers arriving with the animals and style of life they had known at home, Australia was an altered place; the soft native grasses had disappeared. The sheep and cattle trampled and ate them, and in their wake tough, spiny crabgrasses that thrived in the newly hardened soil sprang up. But the sheep and cattle were only the beginning. In 1859, Thomas Austin, in a single instant of nostalgia, unleashed a scourge on Australia from which the country has never recovered. Lonely for home, Austin longed for the fun of the hunt, so he imported and released 24 English rabbits into the Australian countryside. The rabbits bred like rabbits and within 10 years numbered in the millions on each property. One landowner, who in 1860 had a man arrested for poaching rabbits on his property, in 1863 hired 100 men to rid his land of the beasts. The task was to prove impossible. In this rich country benignly free of predators, the rabbits ran wild, changing the landscape of Australia forever. Nothing could withstand their sheer numbers: they stripped the native shrubbery. Like hordes of locusts they began to eat their way across the continent, devouring undergrowth that had sustained a multitude of native animals.

In a desperate bid to contain the plague, Australians drew on the only technology they could. They built great wire rabbit fences — 120 centimeters (4 ft.) high and 60 centimeters (2 ft.) into the ground — to check the advancing blight, but before construction of the fences was finished, the rabbits were across. It took until the 1950s for a solution to be found. When the virus myxomatosis was introduced to Australia, 99 percent of the rabbits succumbed — but even that wasn't enough. The 1 percent remaining developed an immunity, and today Australians are desperately trying, through genetic engineering, to develop stronger viruses in the lab to counter the resurgent millions.

As bad as they were, rabbits weren't the end. Practically every animal that was introduced to Australia over the past 200 years ran wild — pigs, donkeys, camels, to name a few. And plants such as the prickly pear, brought to create English gardens in the 1800s, overran the country in the 1900s. Through it all, Australians never learned the lessons of the ecosystem. In the 1930s, worried about insects devastating the sugar-cane crop, someone came up with the bright idea of importing the South American cane toad. No ordinary toad, this half-kilo (1 lb.) dinner-plate-size solution became a heavyweight problem in itself. The sugar-cane beetles flew. The toads stayed on the ground. The two did not meet and both flourished. Today the cane toad has occupied 500 000 square kilometers (193 065 sq. mi.) of the state of Queensland, and nothing can stop it.

Time and again over those fateful 200 years, Australians have introduced a biological disaster. Today, Australia is a land transformed, or as the less charitable see it, grotesquely disfigured. The land has been overgrazed by 160 million sheep, 14 percent of all the sheep in the world. The precious topsoil is eroding 50 times faster today than it did before 1788. And Australians find themselves frantically trying to regreen a scorched earth.

Australia's story is in many ways universal. Out of ignorance, or in naive attempts to manage an unknown ecosystem, we have, often with the best of intentions, tipped the delicate natural balance. Faced with the great tapestry of life, we seem compelled to unravel it and remake it to portray our wilful and narrow world-view. Often we cast ourselves as fixers, and the destruction we cause is inadvertent. Too often, however, we act out of negligent self-interest.

On a desolate island in the Pacific Ocean stands one of the great mysteries of the world. Archeologists speculate about the origin of the bizarre statues that loom over Easter Island. Out in the middle of nowhere, about 3200 kilometers (1988 mi.) west of Chile, on an uninhabited island, are a thousand statues, each 18 000 kilograms (18 tons) of volcanic tufa standing up to 4.5 meters (15 ft.) tall. Who carved them, and why? We may never know. The civilization that created them is gone, lost in history, but an English botanist has found that it may have been the architect of its own destruction.

The clue lies in what one finds on the island today. Just as strange as the massive statues that stand watch there is the fact that Easter Island, unlike its Pacific neighbors, is treeless. That wasn't always the case. From fossil pollen and some ancient fruit found on the island, botanist John Flenley deduced that a fruit palm had flourished there for thousands of years. By tracking the pollen record, he was able to pinpoint the decline of the tree species as beginning about 1200 years ago and continuing for several hundred years until the tree became extinct. What cataclysm occurred that ended the millennia-long period during which the fruit palm thrived? The arrival of man.

Flenley believes that these ancient carvers arrived on the island and began to clear land for agriculture; they felled the trees to make canoes and levers for raising the statues, and for firewood; as many as 20 000 people were overusing what should have been a renewable resource. What nags at Flenley is that they were doing in microcosm exactly what we are doing to the planet today. But what nags at Flenley most is a terrible image of mindless destruction. Easter Island is so small that "if you stand on the peak of it, you can see almost all of it," says Flenley. "The man who felled the last piece of forest knew that he was felling the last piece of forest, but he did it anyway. What could he possibly have been thinking as he looked around the island, then cut down the very thing he depended on to survive?"

In 1990 the spirit of that Easter Islander hovers over the globe. And now we must ask ourselves, what can possibly be going through our minds as we threaten to destroy the very things we depend on to survive. On the western coast of North America stand the remnants of a once-magnificent temperate rain forest. Stretching from Alaska to California, these trees can trace their ancestry back more than 10 000 years. In the Carmanah Valley in British Columbia, one old giant, a Sitka spruce 90 meters (300 ft.) high and 3 meters (10 ft.) in diameter — perhaps the largest one living — towers over the surrounding trees. It began its life more than 500 years ago, long before the modern chain saw that now threatens its relatives, long before the Industrial Revolution, before a white man stepped onto the shores of North America. All along the coast, these trees are in danger, targeted to be cut — to create jobs and profit and to fill the gaping maw of corporate greed. The coastal rain forest of British Columbia is given 15 years before it is wiped out from logging. In Alaska's Tongass forest, the Sitka spruce were once prized as mainmasts for the clippers that cruised the world. Now you can buy one for the price of a cheeseburger. The U.S. forest service sells them for $1.60 apiece, mainly to Japan as pulp. Our natural heritage is being destroyed for short-term profit and short-term gain, some of it so quickly that we almost didn't even know it ever existed.

That's the way it was for Steller's sea cow. This 9 meter- (30 ft.) long, 3200 kilogram (3.1 ton) marine mammal lived unknown to man on one tiny group of islands in the northernmost Pacific Ocean. It flourished on the Commander Islands of Russia in great numbers until November 4, 1741, when the Dane Vitus Bering explored the area. The sea cow was named for the only naturalist who ever saw it alive, a German named Georg Wilhelm Steller, who wrote this account of it:

These animals love shallow and sandy places along the seashore. With the rising tide they come in so close to the shore that not only did I on many occasions prod them with a pole but sometimes even stroked their backs with my hands. Usually entire families keep together, male and female, long-grown offspring and the little tender ones. They seem to have slight concern for their life and security, so that when you pass in the very midst of them with a boat, you can single out the one you wish to hook. When an animal caught on a hook began to move about somewhat violently, those nearest in the herd began to stir also and attempted to bring succor. To this end some of them tried to upset the boat with their backs while others tried to break the rope or strove to remove the hook from the wound by blows of their tails. It was a most remarkable proof of their conjugal affection that a male, having tried with all his might, but in vain, to free the female caught by a hook, and in spite of the beatings we gave him, nevertheless followed her to the shore, and that several times, even after she was dead, he shot unexpectedly up to her like a speeding arrow. Even early the next morning when we came to cut up the meat to bring it to the dugout, we found the male again near the female's body, and the same thing I observed on the third day, when I went up there myself for the sole purpose of examining the intestines.

Twenty-seven years after Steller's sea cow was discovered, the last one was butchered.

Today we live in an era of "last ones." We're told that elephants, rhinos, and tigers will exist only in zoos within our children's lifetime. Three hundred African elephants a day are killed — for trinkets. That senseless slaughter has gone on and will go on daily until all the elephants are sacrificed. In eight brief years, Africa's elephant population has been halved through poaching from 1.2 million to just over 600 000 — all to feed our consumer craving for ivory. The best projections say the elephants may last until the year 2030. The pessimistic scenario dates their demise at about 2010.

We create international laws like CITES (the Convention on International Trade in Endangered Species) to try to protect endangered species from ourselves. We mutilate animals on the verge of extinction in order to save them from our greed — for example, by cutting off the coveted horns of African rhinos so that the animals are no longer attractive to poachers. It is the unending demand for land for human use that destroys habitats and the combined forces of poverty and greed that fuel the relentless slaughter. What we don't destroy we seek to control. And through all of this we cling to the belief that science and technology are the tools that will allow us to manage the planet effectively.

Science, it has come to be believed, provides us with the knowledge and the means to transform the planet into what we need, or simply want it to be. And when Science hesitates and stops to ponder, as Francis Bacon (1561–1626) did when he wrote that nature, to be commanded, must be obeyed, we dismissively call it Philosophy and plunge ahead anyway, driven by the momentum of past achievements, current needs, and a future guaranteed by our collective adherence to a body of *sacred truths*.

It is said that more than 90 percent of all scientists who have ever lived are still alive and publishing today. Twentieth-century science, especially since the end of the Second World War, has mushroomed, and science, when coopted by industry, medicine, and the military, has revolutionized every aspect of the way we live. An explosion in technological wizardry to aid scientists has created a sense that the knowledge gained permits us to understand and control all of nature. Experts in management of salmon, forests, and wildlife, and enforcers of quality control of air, water, and soil foster the illusion that we are capable, and entitled, to interfere in natural process.

In fact, that is a terrible delusion that anyone who understands the nature of scientific inquiry and insight understands. Scientific knowledge is fundamentally different from other ways of knowing and describing the world. A "worldview" is the sum total of a culture's insight, experience, and speculation. It is an integrated body of knowledge that incorporates values and beliefs as well as profound observational material in a holistic manner. Thus, in a worldview, the planets, a river, mountains, rocks, and so on are all interconnected, as are the past, present, and future.

Scientists examine the world in a very different way. The heart of the scientific approach is to focus on a single part of nature, whether a star, a plant, or an atom, and to try to separate and isolate it from all else. If one can bring it into the lab where it is carefully controlled, all the better. By controlling everything impinging on that fragment of nature and measuring everything within it, a scientist acquires knowledge — about that isolated fragment. But that information does not tell us about the whole, or about the interactions between the parts. Yet scientists today, with few exceptions, continue to examine the natural world in isolated fragments on the assumption that a sufficiently large inventory of pieces will yield a complete description. The picture we acquire is a vague sketch of the complex diversity that exists in nature.

Even if the reductionist assumptions were correct, it is the height of arrogance to think we are approaching a level of knowledge that enables us to manage a natural resource. Scientists don't even know the number of species on Earth because so little research money has been devoted to acquiring such a catalog. Rain-forest biologist E. O. Wilson puts it more graphically: "More money is spent in the bars of New York City in two weeks than is spent annually on tropical research." And even those who have devoted their lives to such study are severely hampered by the technical difficulties.

We know in detail about the basic life cycle and biology of only a handful of plants and animals, and our understanding of the interaction of species in a diverse community is neglible. A decade ago, marine biologists based their understanding of ocean food chains and energy flow on plankton, the tiny plants and animals that can be trapped in fine-mesh nets. Then, *picoplankton*, single-celled organisms so small that they pass through the finest nets, were discovered. Today, picoplankton are thought to be so numerous that some scientists believe they are a major source of atmospheric oxygen. Yet 10 years ago, we didn't even know they existed.

Perhaps one of the species most extensively studied has been the fruit fly *Drosophila melanogaster*. The center of attention for geneticists for 80 years, the fruit fly has cost hundreds of billions of dollars in research funds, consumed millions of person-years, and earned several Nobel Prizes. Yet even the most fundamental questions — how the insects develop from an egg through larva to adult — remain mysteries. It is astonishing, therefore, to hear foresters or fisheries experts confidently speak of managing natural "resources" about which they are far more ignorant than those fruit-fly geneticists are about *Drosophila*.

Science has not provided, and will not provide in the foreseeable future, the knowledge that we need to dominate and control nature. Every practicing scientist quickly realizes the enormity of our ignorance and that it is a perversion of the scientific enterprise to turn the small gains made in the past few decades into major triumphs of human knowledge. But that has not stopped us from doing so.

Inherent in our human conviction that we can manage the planet are ideas that seem so basic and true to societies around the world that they are never questioned — nature is infinite; the biblical injunction to go forth and multiply and dominate the Earth is the human mandate; pollution is the price of progress; growth is progress; all of nature is at our disposal. Yet it is these *sacred truths* that not only blind us to the reality of the environmental crisis but are the cause of it. We don't see that our current beliefs and values are right now compromising the very systems that keep us alive. Like Sampson blinded, we are straining at the pillars of life and bringing them down around us.

PART II
SACRED TRUTHS

THREE

NATURE IS INFINITE

After the tragic fire in 1986 at the Chernobyl nuclear plant, radioisotopes were detected over Sweden within minutes and over Canada's Arctic in hours. It was a grim reminder that the air is a global commodity, a planetary system. No one nation has its own supply of air any more than any one of us has a private stock of it. The air of the Earth is finite and shared by all life forms. We take atoms and molecules from the air into our lungs, absorb them, and incorporate them into our bodies; we give back some of our own as we exhale carbon dioxide. As the scholar and author Jacob Bronowski said: "Sooner or later every one of us breathes an atom that has been breathed by anyone you can think of who has lived before us — Michelangelo or George Washington or Moses." The less comforting reality is that the 12 000 breaths the average person takes in a day are made up of the total accumulation of the world's atmospheric detritus.

Every schoolchild encounters the water cycle — ocean water evaporates, forms clouds, falls as rain into rivers and lakes and oceans, and the cycle begins again. Just as air links all forms of life in interdependence, so, too, does water. The water that we drink in Canada was once transpired from the canopy of the Amazon rain forest or evaporated from the steppes of the Soviet Union.

And the soil provides us with food. All the nutrients necessary for our survival come from plants, animals, or microorganisms — which extract their nutrition from soil and require air and water as we do. Since long before the fifth century B.C. when Empedocles first conceived of the four elements — earth, air, and water, cycled by the fire of the sun's energy — they have sustained the planet and its life forms.

With all that in mind, it seems nothing short of madness for us to use air, water, and soil — the very support systems for *all* life on Earth — as a sewer for our industrial waste. We can't live for more than a few minutes without air, a few days without water, a few weeks without food. Yet we continue to destroy the very things that keep us alive and in the process rob our children of a future.

What propels us on this lemming-like rush to the brink of species suicide? A belief in an endless nature with an infinite capacity to process our excreta and renew itself. For most of human existence our species was nourished by the abundance of the biosphere. As hunter-gatherers, we lived in small family groups, and the impact we had on our natural surroundings was slight. Nature seemed vast, self-sustaining, and perpetually able to replenish itself. Forms of human habitations and social groups changed over time, becoming more complex and more demanding, but we have stubbornly clung to that original view of nature and passed it on, generation after generation, as a form of received wisdom.

Yet conversation with our "elders," citizens of Earth who have lived in one place for their lifetime, reveals that the planet is, in fact, mutable, that the world has changed over the past 70 or 80 years, beyond recognition. Maritime fishermen, Quebec sugar-bush owners, shore-dwellers along the Great Lakes system, prairie wetland hunters, British Columbia loggers — all have seen the limits of what we are determined to believe is infinite. In cities around the world, every day, with every breath we take, we are paying a price for our belief in the "infinite." We love our children and want the best for them. Yet witness Mexico City — it is a city that is poisoning its children.

On a January day in 1986, hundreds of migrating birds fell out of the sky over that smog-choked city. Although no one directly linked the birds' deaths to air pollution, autopsies revealed that their blood contained high levels of lead and cadmium. That came as no surprise.

Mexico City is the most populous urban area on Earth. Two million people lived there in 1940; today more than 20 million people live and work and bear their children in the most poisonous air on the planet. The pollution is so bad that many environmentalists claim Mexico City could become uninhabitable in this decade.

Every day in Mexico City, three million cars and 39 000 factories spew a deadly brew of chemicals into the air: that's 45 million tons of contaminants each year being breathed and eaten by, and stealing years from, every man, woman, and child who lives under the city's blanket of man-made death. Pollution reached such high levels in January 1989 that the local government, fearing for the health of children, closed all schools for a month. That same year a U.N. Environment Program report stated that the ground-level ozone levels in Mexico City exceeded World Health Organization standards by 60 percent.

There are two kinds of ozone: "good" ozone exists naturally in the upper atmosphere and protects Earth from ultraviolet radiation; "bad" ozone is formed near ground level by chemical reactions occurring in emissions from industrial plants and automobiles. It is the main component of urban smog. This ozone can impair lung function and increase the risk of chronic heart disease, asthma, bronchitis, and emphysema.

Respiratory and intestinal infections are the primary killers of Mexico City's children. Some countries, including Canada, advise their diplomatic personnel to avoid taking their children with them to Mexico's capital and not to consider having children while posted there. One out of every 100 children born in this city suffers some form of mental retardation, says one of Mexico's leading geneticists, Antonio Velasquez Arellano.

That horrible statistic is attributable to lead emissions. Fetal exposure to levels of lead once considered safe is now linked to impairment of mental development in infants. Four tons of lead are deposited into the city's smog cover daily; four times as much lead shows up in residents' blood levels as in those living in Tokyo and twice as much as in Baltimore's population.

For unborn children, those levels mean tragedy. A 1987 study showed that under some conditions fetal lead levels are higher than the pregnant mother's. The fetus may even draw off the excesses of lead from its mother's bloodstream. In a study done in Mexico City, seven out of 10 newborns showed lead levels higher than World Health Organization standards.

In 1989, the Mexican government attempted to come to grips with their city's pollution crisis by instituting mandatory testing for emission levels and limiting the use of private vehicles to six days a week. By 1993, all new cars manufactured in Mexico will be required to have catalytic converters. But the efforts are a drop in the bucket. New cars represent only 5 percent of the vehicles on the road each year.

The scope of Mexico City's problem is daunting. Even the most stringent of anti-pollution efforts undertaken immediately would produce only a "slight improvement in 10 years' time," reports a recent U.S. Embassy study. What is even more daunting is that Mexico City's problem is the world's.

In the first comprehensive global assessment of our

toxic cities, a 1988 U.N. survey estimated that 1.25 billion urban dwellers live with unacceptable levels of air pollution. The global village has been contaminated. In the East German city of Bittefeld, 90 to 100 percent of children suffer from respiratory disease. In Athens, death rates are six times higher on heavily polluted days than on those when the air is relatively clear. In Hungary, the National Institute of Public Health concluded that "every 24th disability and every 17th death . . . is caused by air pollution." The U.S. Environmental Protection Agency's 1989 annual report concluded that more than 100 million Americans live in areas where the air pollutants exceed federal standards. The EPA data revealed that some 1.1 billion kilograms (2.4 billion lb.) of toxic chemicals are released into the air each year by chemical-manufacturing plants and other industrial sources. That's more than 320 chemicals, 60 of which are listed by the U.S. government as causing cancer. Ninety percent of all Americans carry measurable concentrations of dioxins, furans, chlorobenzene, dichlorobenzene, benzene and styrene in their fatty tissues, according to the EPA. All six chemicals are thought to be human carcinogens.

And where does Canada sit in all this? As one forthright newspaper reporter delicately put it, "Our record stinks on air pollution." Canadians pump more toxic pollution into the air per capita than do most citizens of leading industrial nations, according to the Paris-based Organization for Economic Cooperation and Development (OECD) — more sulfur dioxide, more carbon monoxide, more particulates such as soot than do Americans, the French, the West Germans, Italians, British, and Japanese. And we rank second only to the Americans in per capita emission of nitrous oxide. This is a significant greenhouse gas and also an ozone depleter. Nitrous oxide occurs naturally, but is also produced by burning fossil fuel and is released into the atmosphere from fertilizers.

We all have a right to breathe fresh air; after all, it is fundamental to life that air be clean. However, as well as a right to clean air, we all have a responsibility to stop polluting it. It is a global problem, requiring a worldwide solution, and no amount of finger-pointing or economics-inspired compromise or mulish insistence that nature is boundless will stop children from dying in Mexico City, or Budapest, or Athens, or in any downtown urban hospital.

Domes of polluted air hovering like poisonous mushrooms over our cities are a phenomenon that has been with us since the Industrial Revolution. Pollution was once considered to be a local problem, the price a region paid for development. That is no longer the case. Because we all breathe the same air on the planet, whoever pollutes the atmosphere pollutes it for the whole world. "The storm is going to come in the 1990s," warns consumer advocate Ralph Nader. "It's one thing to have a river that's polluted here, and an air-pollution inversion in a city there. But now the new ecological spectacles are global: they are acid rain, they are the greenhouse effect, they are the ozone hole. It's almost like an invasion from Mars."

In fact, the problem is far more insidious than an invasion of aliens. These pods that clone destruction are of our own devising. The air carries destruction to all parts of the globe — destruction like acid rain.

Acid rain is not an act of nature; it's the byproduct of industry, a man-made gift of death from the United States to Canada, from Great Britain to Norway, and from Germany's industrial heartland to its farthest reaches. We cannot defend against this deadly rain. Everywhere it falls, it destroys forests, lakes, agricultural lands, washing away human lives and the dreams on which they're built.

Two gases — sulfur dioxide and nitrogen oxide — are the main culprits. These gases turn into an acid solution in the atmosphere and fall to the ground as rain or form mist or fog. Once in the soil, this chemical cocktail converts the pool of nutrients into a toxic metal chaser. The lethal drink is slurped up by trees, and the effects are dramatic. Leaves turn color prematurely, in summer; young crown branches die off; bark peels and the core of the tree rots. Acid rain, in effect, eats away at a tree's immune system, leaving the tree defenseless against natural attackers.

We've caused, then aided and abetted, the scourge every step of the way. Contaminants that fall to Earth as acid rain have been accumulating in our soils since the Industrial Revolution. The buildup accelerated in the 1950s with the proliferation of oil refineries, factory chimneys, and fossil-fuel-burning power plants. Cities and towns became choked with smog from industry and cars. In the late 1960s, industry raised its smokestacks to lofty heights — 150 meters (500 ft.), then more than 300 meters (1000 ft.) — to keep the offending pollutants out of our own backyards. That made the long-range transport of acid rain possible, creating truly global desecration. These poisonous byproducts of our economic progress didn't disappear into thin air as we thought they would.

We first noticed the damage in North America in the heart of maple-syrup country. The blight is devastating maple bushes in Vermont and Ontario. The situation in Quebec's forests is already disastrous. More than half the maples are dead or damaged. Maple-syrup producers are not only losing their cash crop, but mourning the end of a centuries-old tradition. Some have started to save part of their inventory each season for posterity. They fear their children will have to read about the *cabane à sucre* in history books. Overwhelmed by a terrifying sense of powerlessness, they can only stare in disbelief at the sorry remnants of their once world-renowned maple stands.

Michael Herman's majestic sugar bush in Quebec's Eastern Townships was one of the first to succumb to the ravages of acid rain. More than 60 percent of his trees are a pulpy mess. The rest are hastening toward the same fate. Witnessing this relentless destruction has frightened him. "If my trees are dying because the air is so bad, imagine what it is doing to our children."

Acid rain forced Herman to shut down his own operation. For 15 years he has been a broker of maple products, buying directly from farmers who are trying to hang on to their livelihood and their heritage. Every day he sees the desperation in their eyes: "A man comes in with low-quality maple syrup, and less of it, has his arm around his 12-year-old boy, and says: 'What am I going to do? My grandfather gave the sugar bush to my dad and my dad gave it to me. What am I going to give to my son? A goddamn pile of firewood?'" Herman chokes back his tears as he recalls this story, one among many.

The very symbol of Canada — the maple leaf — is being strangled by airborne pollution spewing from industry and car exhausts on both sides of the border. Even though scientists have been vocal about the effects of acid rain for more than a decade, the political will to deal with the issue was markedly absent for much of that time. Our federal government was busily pointing the finger of blame at the United States until 1985 when we owned up to the fact that our own pollution was causing damage. That belated admission of guilt led to a commitment to cut sulfur dioxide emissions by half by 1994.

Although that hurdle may have been cleared at home, there was still foot-dragging in the United States. The powerful coal-industry lobby wielded much influence with former president Reagan, who once maintained that "trees cause pollution." Finally, in 1986, he had to acknowledge that acid rain was a man-made threat.

Even so, as late as 1989, U.S. Justice Department lawyers persisted in arguing that American industry's responsibility for Canada's acid-rain problem was "entirely speculative." In 1990, the Bush administration cleared the way for a new Clean Air Act, which includes the first-ever acid-rain controls — a 50 percent reduction in acid-rain-causing emissions by the end of the decade.

While Canadian politicians applaud this legislation as a major breakthrough (it has now opened the door for a U.S.–Canada Acid Rain Accord), scientists feel that a 50 percent cut in emissions is too little, too late.

"That's only going to slow down the damage. It won't eliminate it. It would take 25 years or more to see any signs of recovery, even if we turned off all the pollution taps tomorrow," says McGill University forestry expert Arch Jones.

Time is running out faster than we think. Acid rain kills off its victims with alarming speed. Even the Worldwatch Institute, which prides itself on being the pulse-taker of the world, was caught off guard. While writing the first "checkup" report, *State of the World 1984*, the institute's president, Lester Brown, debated whether to include acid rain in his coverage in light of the findings of German foresters that 8 percent of their nation's forests were showing signs of damage. Just six years later, more than half of Germany's forests are dead or dying.

This *Waldsterben*, or forest death, in Germany's Black Forest gave the world its first glimpse of the acid-rain menace about 10 years ago. Regal spruce and fir trees that had flourished for centuries suddenly and unexpectedly began to shed needles and turn brown. The death toll has been mounting at a dizzying rate ever since.

The German biochemist Bernhardt Ulrich, who first predicted the death of the forests, says 90 percent of the country's trees will be gone by early next century. The lush green blanket now covering a third of

Germany will be reduced to bleak barrens pitted with moss and bogs.

The loss cuts to the very core of national sentiment. Germans value the outdoors and the beauty of nature so much that they forbid the posting of Keep Out signs near any forest. But Germans also treasure their right to drive at unlimited speeds on the autobahns. Some would say the car has usurped that special place in the German soul reserved for the love of the forest. As a result, fumes from 20 million car exhausts have wiped out the competition — the forests — for pride of place. Add to that the deadly pollutants spouting from power plants and factories in the Ruhr Valley and the real tragedy becomes clear: Germany is systematically destroying its most cherished natural resource.

It is neither a national nor a continental cataclysm. Acid rain knows no boundaries. It's estimated that eight million hectares (20 million acres) of forests throughout Europe are suffering from acid-rain pollution, and the deadly deluge has been reported in places as far-flung as China and Africa. Trees are not the only victims.

Acid rain is a killer of all forms of life on the planet. Our infected soils are silently leaching their toxins into our supply of fresh water. The acidic runoff carries with it dissolved elements that become the poisons dropping into our lakes, rivers, and streams, eliminating fish and plants. The Canadian department of Fisheries and Oceans estimates that 14 000 lakes, stretching from Manitoba to Newfoundland, are already dead, and 150 000 more are at risk.

The most vivid evidence of acid-rain contamination of the aquasystem can be found in Nova Scotia. Salmon have disappeared from 13 rivers. The fish suffocate as they gasp for air and salts through gills clogged with a toxic aluminum mucus. Scientists are trying to halt the destruction by adding buffering agents, such as lime, to the water. Some lakes have recovered, but liming is only a stopgap, not a solution. The only way to permanently reduce the acidic content of the water is to identify and eliminate the cause of the problem. And the cause lies in our stubborn adherence to the *truth* of a nature without limits. It may be a fatal flaw. As any photographer knows, setting the focus on infinity blurs the foreground. By viewing the world as infinite we simply cannot see that our local actions have global consequences.

Yet no place on Earth is exempt from the inexhaustible variety of pollutants we create. We have lived with the understanding that we could dump them into the air and the water and somehow they would just go away. But the reality is that they are carried off by prevailing winds and rushing ocean currents to destinations we would never have suspected.

The eastern coastline of Hudson Bay is dotted with small, isolated communities. The Inuit who live there hunt, fish, and enjoy the bounty of this northern land in much the same way their ancestors have done for thousands of years. Civilization has barely encroached on their way of life. The bark of the husky is the most prevalent sound, not the roar of car engines. No smokestacks mar the stark landscape of snow and ice that seems to stretch to any imaginable horizon and beyond. Anyone would think that these people inhabited the cleanest environment on Earth.

A group of scientists at Laval University's Centre Hospital certainly believed that to be true. That's why they selected 200 Inuit women from these communities to participate in a 1987 study aimed at measuring contaminants suspected to be in the breast milk of women living in the highly industrialized area around Baie-Comeau and Quebec City. The Inuit women were thought to be contaminant free — the perfect control group, the most pristine point of comparison to be found anywhere.

When the scientists looked at the preliminary results in the lab, they were shocked. The contamination levels were certainly high — but in the breast-milk samples from the Inuit women. The toxic chemicals included levels of PCBs five times higher than those discovered in the other samples. Some chemical levels were higher than those recorded anywhere else in the world.

What that study hammered home is that the pristine Canadian North sits at the center of a deadly vortex of swirling poisons discarded by the rest of the world. "The outpourings of the great rivers of the Soviet Union flow into common Arctic waters and undoubtedly contribute PCBs, now banned, but originally used in electrical transformers. There are airborne chemicals from insecticides sprayed on cotton fields in the southern United States and from DDT used in Africa that are showing up in the livers of fish in the Mackenzie River, and in seal, in walrus, and in polar bear on Baffin Island," says Dennis Patterson, government leader of the Northwest Territories.

These same chemicals, along with other pollutants, have been found in plankton off the coast of Ellesmere Island in the High Arctic. Radioactive cesium, first spotted in lichen eaten by caribou in the 1960s when the North was used as a nuclear-weapons-testing site, has reached dangerous levels in caribou herds since the 1986 nuclear reactor disaster at Chernobyl. People living in northern Quebec have been warned not to eat the livers and kidneys of deer and moose because of cadmium contamination.

Scientists have concluded that northern peoples are being poisoned by their heavy diet of polluted fish, seal, whale, and walrus. Man-made chemicals such as PCBs accumulate through the food chain, from plankton to fish, and then concentrate in the fat of marine mammals. Blubber, in particular, is a food staple of the

Inuit. "The very survival of our people is at stake," says Patterson. The North's 22 000 Inuit, along with their 13 000 western neighbors, the Dene and Métis, have the rest of the world to blame if they are now forced to join the growing ranks of environmental refugees.

Giving up their traditional diet is more than impractical for the Northern peoples; it would result in the annihilation of an entire people, says Lorene Gemmill, the regional nutritionist for the Baffin Region of the Eastern Arctic. It represents "the destruction of their culture. Food is a very important aspect of Inuit life. If, all of a sudden, that is taken away from them, you're changing a lot more than just their diet. We'd lose a lot of the Inuit and we'd have a big gap in the heritage of Canada."

On a tiny island just off the east coast of Baffin Island, a community of some 400 Inuit are already living with the fear that such a loss creates. For thousands of years, the biggest threat this community had to face was a charging polar bear or a raging blizzard whiting out the sky. Today, an enemy they cannot taste or feel or see is defeating them and putting their lives in jeopardy. On Broughton Island, where more sea mammals per capita are consumed than anywhere else in the North, one out of five residents has unacceptably high levels of PCBs in his or her body, according to a federal government study of Arctic contaminants conducted from 1985 to 1988.

Pauloosie Kooneeloosie is a hunter of almost legendary fame in the community. He is the father of 10 children. He has never smoked or touched alcohol. He has never known anything other than hunting and living off the land. His lifestyle is his identity. Now, he is being told that he is being poisoned by it, that the only world he knows is being destroyed by pollution from industrial nations on the other side of the world. "Long time ago, we never had to worry about any kind of poison in the food. Starting back in the 1970s we began hearing about such things. Now southern scientists are telling me the poison is in the seal. It worries me," he says in his native Inuktitut.

For the people of the North, the land is the provider of food, clothing, and shelter. The harsh northern environment is a constant reminder that the natural world is not to be trifled with or taken for granted. "It is ironic that the people of the North should be the first to suffer the direct consequences of this disrespect, considering their relationship with the land. But maybe that is needed to smarten up the citizens of the world," observes Mary Kaye May of the Kativik Regional Board of Health and Social services.

The children of the North are painfully aware of the future they will inherit if the citizens of the world don't smarten up. Schoolchildren in Iqaluit, Baffin Island, compose essays these days with such titles as "My Polluted World." For many kids, today's reality is frightening enough: "I could get sick, possibly ending in death. I never thought the Arctic would be polluted because it's so clean and there are no factories up here."

Industrial effluents but no industry; PCBs but no electrical transformers; DDT but no crop pests; cesium but no nuclear-power generators — the myth of the pristine North is concrete evidence that the infinity we've come to depend on may be measurable in decades, in our children's lifetime. For 99 percent of human history the degradation of the natural world seemed to us to be invisible, at worst reparable. Our numbers were small, and we could always move on. But for the 5.3 billion of us here today in 1990 there is nowhere left to move on to, and the reality of our situation comes as a shock. We live in a *finite* world. There is no magical power of limitless regeneration, no endless expanse of air, soil, or water. Not a spot on the globe remains free of our pollution.

Just as we have the air, we have used the waters of the world — one of our primary food sources — as a fathomless garbage pit. Now, in a thousand and one ways, they are telling us we have pushed them to the limit, in some cases, too far.

As we enter the last decade of this century, the oceans, seas, lakes, and rivers of the world read like a roll-call of the dead and dying: scientists predict that the North Sea will be dead within the next five years and it will not be death by misadventure or natural causes, but murder. For 1000 years, the North Sea has absorbed the filth of Europe's cities and, for 1000 years, it seemed to cleanse itself. Now, the assault has become overpowering. Each year European nations dump 500 000 tons of nitrogen, 46 000 tons of phosphorous, 110 tons of mercury, 1100 tons of cadmium, 21 800 tons of lead, and 107 100 tons of zinc into North Sea waters. In the summer of 1988, 10 000 seals in those waters died agonizing deaths, killed by a virus that scientists say their pollution-weakened immune systems could not withstand.

The Atlantic Ocean is turning into a deadly chemical soup for its inhabitants. Since June 1987, at least 750 dolphins have washed ashore along the Atlantic coast of North America, many of them bloated with toxic chemicals, their flippers, snouts, and tails blistered and raw.

Coastal cities have turned ocean waters everywhere into cesspools of human waste. Three million people live within a 40 kilometer (25 mi.) radius of Boston harbor, and nearly all the human and industrial waste of the metropolitan area winds up in its waters.

Vancouver's sewage washes directly back onto local beaches, and is used for agricultural irrigation. A 1989 Greenpeace study reported that Vancouver residents swim in and eat their own poisons. In one British Columbia harbor 60 percent of the English sole were found to have tumors. The report called the Fraser River the "biggest sewer line" in British Columbia. The river has become the largest single source of pollution for the waters of Georgia Strait, part of the famed Inside Passage to Alaska.

And what about Victoria's dirty little secret? The elegant dowager of the west coast, British Columbia's charming and neatly groomed capital, has been using the Pacific Ocean not just as her playground but as her toilet. No muss, no fuss, no bother — everything flushed goes straight, in its own natural state, out to sea, through two massive pipes that jut 1200 to 1900 meters (3900 to 6200 ft.) into the ocean. Victoria is one of the few major cities in North America without a sewage-treatment system. The garbage plume makes its way back to shore at least 30 percent of the time, as a casual stroll along Victoria's beaches will attest. In a kind of perverse symmetry, Halifax, the capital of Nova Scotia, on Canada's east coast, pours its untreated sewage into the Atlantic.

Like a weakening patient battling a persistent disease, the oceans are now falling victim to opportunistic infections. Water surfaces everywhere are beginning to carry a deadly blight. All over the world, blooms of algae, sometimes known as "red tide," commemorate our long-held belief that the solution to our escalating problems with garbage and waste was to let the water carry it away. Some scientists blame the blooms on agricultural runoff of fertilizers and pesticides; others point at acid rain or sewage. Whatever the cause, tiny algae or phytoplankton that have lived in equilibrium in the oceans for three billion years are now out of control. As scientists document the spread of the algae worldwide, they are convinced that we have created a monster.

In the summer of 1989, people living along the Adriatic coasts of Italy and Yugoslavia watched in horror as an algae explosion stretching for 640 kilometers (400 mi.) sucked the oxygen from the water and killed tons of clams and mussels. Norway's coastal waters have also suffered massive growths of algae in the past several years. Fish farmers there tried to tow floating pens of salmon and trout away from the infested areas, but they lost more than 609 630 kilograms (600 tons) of fish.

The effects of mankind's monster are reaching right to the top of the food chain. Mussels poisoned by algae blooms in Prince Edward Island killed three people and poisoned 150 others. Recently, there were 87 deaths in the Philippines attributed to shellfish poisoned following an algae bloom. Once the blooms start growing, they can't be stopped: none of the Earth's oceans or seas seems to be immune. Hong Kong had never recorded a red-tide bloom before the middle of the 1970s. Now, with dumping of sewage and runoff of fertilizers, the city's harbor is hit by 23 a year.

As the extent of the destruction of our global waters becomes evident, there is an unspoken fear that what we are witnessing may be the elimination of one of the support systems that we depend on to keep us alive. If we have been waiting for a crisis to convince us that something is drastically wrong, we have one, and the signs are at our doorstep.

Beluga whales living in the St. Lawrence River are so glutted with toxic chemicals such as lead, mercury, and PCBs that when a beluga dies the corpse is classified as toxic waste. The Great Lakes are so polluted with more than 1000 toxic compounds that a report issued in October 1989 warned that women not yet past child-bearing age should avoid eating fish from the waters.

Everywhere, the waters of the world are sounding an increasingly desperate SOS. We are destroying this living resource by what we've given and what we take. No one hears the call more clearly than fishermen.

When mariners first came to the Atlantic seaboard they were impressed by the ocean's bounty there. In 1497, John Cabot returned to England with enthusiastic reports of "a sea swarming with fish that could be taken not only in the net but in baskets let down and weighted with a stone." Word of these riches spread to every maritime land, and soon European ships flocked to the New World to reap their share of the seemingly endless supply of fish.

That hunger continued for the next 500 years. We used nets, harpoons, guns, and even bombs to satisfy our insatiable craving. In the process, we finished off the great bluefin tuna for trophies, sushi, and sandwich filling; we slaughtered the seals for their valuable oil and pelts; we gutted the herring for affordable caviar, and dumped their bodies as landfill. We took and took and nature kept giving. We never stopped to think we were taking too much. We believed there would always be plenty more where that came from.

The greed that grew out of our belief in nature's boundless provisions is threatening to empty the seas. Today, we pay a fee to catch a glimpse of the few remaining whales. The famous North Atlantic salmon is dwindling in numbers. Now, the very foundation of the $2 billion east-coast fishery — the northern cod — is in peril. In the 20 years that Brian Walsh has been fishing Bay de Verde, Newfoundland, he's never been this discouraged: "There's no fish out there to catch, and we can't catch what's not there. God knows we try hard enough. I don't know if it can get any worse than it's been in the past few years."

By 1989, the catch was not large enough to keep all the processing plants open. Continued closures could mean the loss of 35 000 jobs in Newfoundland alone. In a province with a 16 percent unemployment rate, the disappearance of the northern cod is a death blow. Many fishing communities have no other reason to exist. Most fishermen are "angry, disgusted and apathetic — all rolled into one," says Walsh. The northern cod is the one stock fishermen always thought they could depend on. Now they feel betrayed by the *sacred truth* that the sea's riches would never run out.

The basis of this "truth" lies in the arrogant conviction that we can actually manage the ocean's resources. In theory, the strategy is to take as many fish as possible while allowing the stock to replenish itself. The reality is we haven't the means, or the will, to get that balance right. "We can't count the fish in the ocean," says Chris Campbell of the Marine Institute of Newfoundland and Labrador. When a farmer loses his crop to drought, he doesn't have to guess what happened. When a fisherman's nets come up empty, he has to wonder if the fish simply went elsewhere or were harvested by someone else.

The truth is we cannot determine exactly what's going on with a huge family of fish in a vast body of water. "It's one population of fish," says Campbell, "whether the fish is caught by a European trawler on the nose of the Grand Banks, or an inshore trap fisherman, or a plumber from St. John's who's out jigging fish on a Saturday afternoon to fill his freezer. Every fish caught is from that same population and all fishing has an impact. We have to be able to control all of it."

In 1977, Canada thought it could get that control of the fishery by extending its management jurisdiction to 200 miles (320 km.) offshore. European trawlers, aided by sophisticated technology, had spent 30 years plundering the cod stock to near extinction. The new nautical zone was supposed to put an end to this massacre. But in our hurry to revitalize a flagging fishing industry, crucial errors were made. A large and hungry Canadian offshore fleet took over where the foreign trawlers had left off. At the same time, Portuguese and Spanish fishermen continued to drag their nets through spawning cod in international waters just outside the 200 mile limit. No one listened to the inshore fishermen who had been saying for years that the numbers of cod and other fish species were declining.

Harold Barrett has been fishing and building trap boats for 58 years in Old Perlican, Newfoundland, as his ancestors had done since they arrived in this small outport in 1812. "A man that's fishin', he's takin' in what's goin' on. But those scientists" — he shakes his head — "I don't believe our Canadian scientists put half enough time on the ocean to see the real thing. Now there's no turbot left. And the cod? Well, when I was fishin', you could go out here, go anywhere around Baccalieu Island, and you'd catch your cod anywhere from two and a half to four feet long. Now you go out, you get 'em 12 inches long, 18 inches — not too often you gets any cod any size."

It wasn't until 1989 that our federal scientists finally figured out they had it wrong. They discovered they had grossly overestimated the size of the northern cod population and that the stock was being fished at twice the rate it should have been. An independent panel of experts called upon to assess the damage concluded that the total allowable catch had to be slashed by 50 percent before the stock had any chance of growing back. In addition, the panel recommended that young fish be left in the ocean and that an extension of the 200 mile limit was needed to curtail foreign fishing on the Continental Shelf. The final report, released in April 1990, stressed the urgency with which these measures should be implemented.

The panel's chairman, Leslie Harris, the president of Memorial University in St. John's, adds this warning: "We are at the edge of a precipice. Our chance of slipping over increases every hour." We could take that plunge sooner than we think.

The minister of Fisheries and Oceans, Bernard Valcourt, bluntly refused to alter the total allowable catch set for 1990, arguing that further cuts in the quota would spell economic disaster for Atlantic Canada.

The real bottom line is: no fish, no fishery. "It's renewable, but it's finite," says Chris Campbell. "We can treat fish in the same way as developing a coal mine, but then we have to accept we only have a five-year fishery and it's all gone."

That mining metaphor has become frighteningly apt in recent years, and not just in the Atlantic fishery.

Fierce international competition in the Pacific has intensified to the extent that we are actually mining these waters of their bounty. We have become extraordinarily efficient at killing fish. Bigger boats can trawl the ocean for months at a time. They are equipped with high-tech fish-finding devices and the most insidious postwar invention of all — the drift net. Environmentalists have dubbed it "the wall of death" and "the marine equivalent of genocide."

Drift-net technology was developed by Japan, which, along with Taiwan and South Korea, uses the nets extensively on the high seas, ostensibly in trawling for squid. But the technology is an indiscriminate killer of every form of marine life large enough to be entangled in the tough nylon netting openings, which snag gills, fins, and tentacles. A single net stretches out underwater like a vast curtain 50 kilometers (30 mi.) wide and 10 meters (32 ft.) deep. At night it is virtually invisible. Not even a dolphin's sonar can detect it.

The death toll is astronomical. North American scientists and environmentalists estimate the nets have killed close to a million sea birds, dolphins, seals, sea turtles, and a wide variety of deep-ocean fish. Their mangled, rotting carcasses are simply thrown overboard.

Worse yet, the carnage has spread far beyond the squid-fishing areas of the North Pacific. About 20 000 tons of West Coast salmon are illegally caught each year in Asian drift nets, according to fisheries experts. The great Atlantic salmon fishery in Ireland has been wiped out by unregulated drift-netting. When the fishing season in the Northern Hemisphere is over, about 160 Japanese and Taiwanese boats head south to Australia and New Zealand in search of albacore tuna. This species is being fished dry and could disappear within the next two years unless the drift nets are banned.

All that represents an enormous bite taken out of the ocean's food chain — perhaps too much for any biological system to recover from.

In December 1989, an agreement was reached at the United Nations to eliminate the use of drift nets in the Pacific by 1992. The resolution, which applies only to the North Pacific, is not legally binding, even on the signatory countries. It will take massive international pressure to put an end to the slaughter on the high seas. But by then it may be too late. Drift-net technology can outlast our will to use it. Pieces of the net break off, nets are abandoned when damaged, and these castoffs, called "ghost nets," rove the oceans like restless spirits, catching and killing without purpose. There is every reason to believe that we will literally run out of fish if we continue to assume that the planet's waters exist strictly for our use, and damn the consequences. "We have some claim as predators," says Leslie Harris of the Northern Cod Review Panel, "but we have to recognize that we share the resource with others who depend on it. The others are all the rest of the living creatures in the system."

One of the great contradictions in human nature is that we value things only when they are scarce. Not even fishermen are exempt from this misperception. "I guess when you live so close to the ocean you take it for granted — you think the fish, the birds, and wildlife will always be there," observes Brian Walsh. "Our problem is we only appreciate the water once the well runs dry."

And the well is doing just that. We are pillaging not only aquatic life forms but water itself. Would you ever have imagined that the world could be running out of water? That is what we're being told is happening, and not just in traditional drought areas but right here at home in North America.

Americans get 60 percent of their drinking water from groundwater sources, called aquifers. The underground permeable saturated zones of rock have held water for thousands of years. We are pumping out this fossil water at enormous rates, far beyond recharge.

U.S. farmers draw 20 billion more gallons of water daily from underground aquifers than can be replenished by rain.

The Ogallala Aquifer, the largest body of underground water in North America, underlies seven or eight states, from South Dakota to Texas. The Ogallala is being pumped out at a rate of many meters a year and is being recharged by rainfall at millimeters a year. The water in the giant aquifer accumulated over several ice ages; it's already running dry in some areas. Parts of West Texas above a depleted part of the Ogallala are dustbowls. And the rest of it is predicted to go dry early in the next century. The decision to *mine* the aquifer to exhaustion in fewer than 50 years was made deliberately, based on the assumption that there is an "infinite" number of tappable water sources.

Twenty-six percent of Canadians (6.2 million people) rely on groundwater for domestic purposes. Thirty-eight percent of all Canadian municipalities rely partly or totally on groundwater. In Prince Edward Island, more than 99 percent of the population is dependent on groundwater. A seasonal shortage of water can affect almost any part of the country. In summer 1987, residents in parts of Newfoundland had to boil water for drinking; others had to travel long distances to obtain water; factories closed and buildings burned down, all because the water supply was inadequate.

All over the world, water for drinking, household use, industry, and agricultural irrigation is running out. The Worldwatch *State of the World Report 1990* has documented the decline.

Both China and India, the most populous countries in the world, are exhausting their underground water stocks. Groundwater levels are falling by up to a meter a year in parts of northern China, an important wheat-growing area. Fifty cities in China face acute water shortages. In parts of India, water tables have fallen by as much as 25 to 30 meters (82 to 98 ft.) in the past decade. Tens of thousands of villages throughout India now face shortages.

With Israel, Jordan, and the West Bank expected to have utilized all renewable sources of water by 1995, the major clashes in the Middle East in the late 1990s could well be over control of dwindling supplies.

What water we haven't siphoned off for our own use will be affected by global warming. As temperatures rise and rainfall patterns shift with the onset of climate change brought about by greenhouse gases, water supplies will increase in some areas and decrease in others.

Using climate models that predict a two-degree Celsius (3.6°F) rise in temperature with an accompanying 10 percent decrease in precipitation, some researchers foresee as much as a 40 to 76 percent drop in water supplies in river basins in the western United States. Those kinds of shortages in the United States could spell trouble for Canada. Some experts predict heavy pressure from the United States in the future to divert Canadian water south of the border. "I think a worst-case scenario would be a situation where Canada gets itself sufficiently in debt, and sufficiently linked with the U.S. economy, that we have no alternative but to accept what would amount to an American request for our water," suggests David Brooks, the author of a Science Council of Canada report on the future of water. A 1986 article in *The Economist* talked of Americans "dreaming up schemes like diverting British Columbia's rivers to American farmers."

What experts are suggesting is that it's time we stopped pretending that water is anything but a scarce commodity. We take water for granted on this continent. We turn on our tap and it's there. Most of us only encounter shortages during those rare long dry periods when we're asked to refrain from sprinkling our lawns and washing our cars. "Limitless" and "free" are probably two words most of us would use to describe the water we use. "Water is not free," says David Brooks. "It's not like going down to the river and dipping your bucket in." We have to start looking at water resources the same way we started to look at energy during the oil crisis of the 1970s. It costs and it exists in limited quantities.

North Americans are the most wasteful water users in the world. While we wash our cars and water our lawns and run our dishwashers, and keep the tap running while we brush our teeth, almost 2 billion people have access only to an inadequate supply of water that is also dangerously contaminated. Yet the United States withdraws more than twice as much water a person than the Soviet Union, and more than five times as much as either China or India. Canada runs a close second: each day Canadians use more than 2000 liters (528 gal.) of water a person for domestic, commercial, agricultural, and industrial purposes.

"We could be cutting our water down by very large amounts," says David Brooks, "and by large amounts I mean by 50 to 60 percent." Such savings can be accomplished through a number of simple devices and techniques. Taps and showerheads can be fitted with conservation devices to cut down wastage without loss of effectiveness. Efficient toilets are available that flush with one-third as much water as do conventional toilets. As Brooks says, the initial cost may be higher, "but they save money all along, and more important, they save all that water."

In terms of water usage, our economics is false. Because we aren't charged for water or pay a negligible amount, we have no incentive to use it efficiently. If we had to pay the real price of what it costs to get water to us, and the environmental cost for the decline in quality once we've used it, we would be more responsible about its use. Researchers have found that in a number of countries — Israel, Canada, Great Britain, and the United States — household water demand drops by 3 to 7 percent when prices rise 10 percent.

The most aggressive water-efficiency program exists in the state of California. In 1990, that state is experiencing its fourth year of drought. The scarcity of water is raising profound questions about whether California can preserve its comfortable way of life or grow the way it has been. Rationing has been a fact of life for some time on the central coast of the state. In Monterey, each household has to cut back 20 percent from normal use or face stiff fines. The seaside city of Santa Barbara is suffering such a desperate shortage that it is considering building a plant to turn seawater into fresh. In 1976, the state adopted a plumbing code intended to save water, and made it part of California's building code that applies to all new housing.

In real terms the cost of water wastage is unrecoverable. In their *State of the World 1990* report, the Worldwatch Institute considers the water situation so critical that it ultimately calls for "societies to adjust human numbers and activities to water's natural limits." That may mean reusing waste water as Israel now does with 35 percent of its water and plans to do with 80 percent of its waste-water flow by the end of the decade. That may mean developing crops that are more salt- and drought-resistant to make do with less water or more saline water. But finally, what it means is that water priorities will have to change — using scarce water to grow thirsty cotton crops in the desert, as is done in Soviet Central Asia, or to irrigate hay for cattle to eat, as is done in the water-short American West, may simply not be possible for much longer, says the Worldwatch report.

Stanford ecologist Paul Ehrlich attributes the degradation of our fundamental life-support systems to the fact that humanity has been triumphant. We have become, within a matter of a few decades, the dominant species on the planet, "and in our triumph are unfortunately the seeds of our destruction." We've built a gigantic population, which is still growing rapidly, and Ehrlich says that we are keeping it going by doing something that we wouldn't even consider doing with our personal finances — we're squandering our inheritance; we're living on capital. "We couldn't possibly support today's 5.3 billion people on income, and when our capital is gone, we're going to be in deep trouble."

Clean air and water are just two examples of the capital we're exhausting; the deep, rich agricultural soil of the planet is another. Soil that is generated on a time scale of inches a millennium is being degraded on a time scale of inches a decade, says Ehrlich. The agricultural system we have evolved to feed a global population that grows by 1.8 percent a year annually loses "25 billion tons of topsoil, the amount of topsoil that underlies all the heartland of Australia." It blows away; it washes away; it becomes salinated; its nutrients are used up.

Calculations done by the Worldwatch Institute suggest that topsoil loss means an annual decrease in grain production of nine million tons. As global population goes up, food production goes down. According to the Food and Agriculture Organization, soil erosion could reduce agricultural production in Africa by 25 percent between 1975 and the year 2000 if conservation measures are not adopted. That means disaster on a continent in which one-quarter of the population — more than 100 million people — do not consume enough food right now to allow an active working life. One-sixth of the total area of China — 1.5 million square kilometers (579 195 sq. mi.) is now damaged by soil erosion. Soil degradation has been called Australia's greatest environmental problem by the country's prime minister; it affects two-thirds of that country's arable land. In 1984 a Canadian Senate committee report stated that soil erosion was costing Canadian farmers $1 billion a year in lost income. Despite Canada's image as a land of plenty, good agricultural land occupies only 5.5 percent of the country's total land area, and what little we have is relentlessly blowing away. In one area of Saskatchewan, it has been estimated that winds erode 5.6 tons of soil a hectare (2.5 acres) yearly.

Ehrlich thinks we just don't realize the direct impact these facts have on our survival because we've lost touch with the way we are supported by nature:

> The average New Yorker, and I suspect the average person in Toronto or Vancouver, thinks their food materializes overnight by magic in supermarkets. I've had many people say to me, "Gee, I just flew across the continent and there's lots of room for additional people because there's lots of empty space out there in Iowa or Saskatchewan." They don't understand that the green stuff in Iowa and Saskatchewan has a very direct relationship to their lives. . . . That's a very serious problem in our culture, and we don't really make the connection in our schools. Instead of saying "See Spot run" in the early grades, we should say, "See the plant grow in the soil."

In the 1990s the connection will be brought home to us with a vengeance. Diminishing water supplies, soil erosion, and predicted climate change are the ingredients for a lethal mix.

The Worldwatch Institute reports that the first casualty in this altered world we are leaving our children will be our food supply. The signs are already visible. "My guess is that the issue that is going to mobilize the world, to galvanize the world to respond to these environmental threats, is probably going to be food shortages and rising food prices," says Lester Brown, of the Worldwatch Institute. "We knew that we couldn't continue to damage our life-support systems without eventually paying a heavy price, but how would we be affected? When would we know we were in trouble?" Brown suggests we now have the answer.

Food scarcity is emerging as the most profound and immediate consequence of global environmental degradation. It is already affecting the welfare of hundreds of millions of people, and will begin to affect North Americans directly in this decade.

The U.S. Department of Agriculture sized up the situation in the first two lines of its October 1989 report: "Two consecutive years of drought-reduced U.S. wheat production, coupled with unusually low world wheat stocks, have renewed attention on the subject of reliability of supply for the world's importing countries."

What is shocking about that statement is how rapidly the food situation turned disastrous. In 1987, economists were worrying about a glut of grain on the world market. At the start of the 1987 harvest, world grain stocks totaled a record 459 million tons, enough to feed the world for 101 days (according to Lester Brown). Today the world is virtually "living hand to mouth from one harvest to another." An 80-day supply is considered the minimum safety net. There is enough grain in storage to feed the world for 60 days.

Just how precarious has the world's food supply become? Current shortages can be traced to a single event, the drought of 1988. A monsoon failure in India in 1987 set the stage by lowering world grain output by 85 million tons. But fast on its heels came the summer of 1988, the season of no rains — not for China, not for America, not for Canada — the summer that record-breaking heat sucked the life out of crops across the continental prairies, and nature knocked the legs out from under the "green revolution."

At the source lay a single alteration in weather patterns. The jet stream that usually carries rainstorms across the west coast into the Midwest did something unusual in 1988: it split in two. One branch headed north to Hudson Bay, and the other south to Mexico, leaving the midlands arid and hot.

What the jet stream left in its wake that summer was the death of the conviction that the North American "bread basket" could be relied on to feed the world. The United States lost 31 percent of its grain crop and the Canadian harvest was equally devastated. The prairies lost about one-third of their grain crop. Fifty percent of Saskatchewan wheat dried to dust. Disheartened and broken by the drought, more than 24 000 Canadian prairie farmers and farm workers abandoned agriculture that year.

For the first time in its history, the United States did not grow enough food to feed itself in 1988. The drought forced it to dip into grain reserves. It had been predicted that 1989 would be a bumper year for world grain; the silos would be refilled and the summer of 1988 forgotten. It didn't happen. As good as the 1989 harvest was, it was 18 million tons short, and already depleted reserves fell even farther. For the third year in a row, the world produced less grain than it ate.

British climatologist Mick Kelly is concerned that the drought of 1988 was a portent of what lies ahead with a climate changed by greenhouse warming. "If we look 10 years hence, one climate model is predicting drought of this nature six times every decade. The drying out of the continental interiors of the United States and parts of Canada is one of the most definite predictions that our numerical models have given us." Some of the land in the U.S. western Great Plains that now produces wheat would revert to range land. The western corn belt would become semi-arid. During a particularly severe drought year, spring wheat production in

Saskatchewan could be down by as much as 75 percent, warns a 1988 Environment Canada report. In a worst-case scenario for the greenhouse effect, the report predicts five- to 10-year periods of severe drought conditions in that province.

If the summer of 1988 provides a glimpse of what harvests will be like, then the days of the North American bread basket may be numbered. Those days are not to be counted in grains of wheat or bushels of corn but in lost human lives.

The track record for feeding the world, even in times of plenty, has been a miserable one. U.N. World Food Council documents show that the number of hungry people increased five times faster in the 1980s than in the previous decade. By 1989, 550 million people filled the ranks of the malnourished or hungry. For these millions, hunger is a harsh reality right now. Every day around the world 40 000 people die of hunger. That's 28 human beings every minute, and three out of four of them are children under the age of five. And that is the toll in times of relative plenty.

Today, food stocks are down, and prices are sharply up. U.S. wheat prices increased by more than 40 percent in 1989. Those who need the food most will be least able to cope. Perhaps a billion or more of the world's people already spend 70 percent of their income on food. For many in this group, a dramatic rise in the cost of grain is life-threatening.

Television has habituated us to the anguish of death by starvation. We've gotten used to the idea that somewhere in the world someone is desperate for food. However deeply hunger may have touched our hearts, it hasn't affected our lives. Now an ominous warning comes from ecologist Paul Ehrlich:

There are some people who seem to have the idea that places like Canada and the United States can remain islands of affluence in a sea of misery. It's just not going to work that way. As people around the world get more hungry and more desperate, they're going to come pounding on our doors. They're not going to sit quietly and starve to death. There's going to be a flood of ecological refugees that's going to be very difficult to stem. In addition, of course, more and more of the poor nations in the world are going to be armed with nuclear weapons, and that will become a gigantic threat to countries like Canada and the United States.

So Ehrlich thinks we can forget about the idea that "we" can live happily ever after while "they" go down the drain. Our fates are intimately tied up with their fates, and "we're either going to save the world, or we're not going to save ourselves. We can't just save ourselves and let them go," says Ehrlich.

Lester Brown's scenario gives a terrifying urgency to Ehrlich's words:

If the United States were to experience a drought-reduced harvest similar to that of 1988, before stocks are rebuilt, its grain exports would slow to a trickle. By September of that fateful year, the more than 100 countries that import U.S. grain would be competing for meager exportable supplies from Argentina, Australia, and France. Fierce competition among importers could double or triple grain prices.

By November, the extent of starvation, food riots, and political instability in the Third World would force governments in affluent industrial societies to consider tapping the only remaining food reserve of any size — the 450 million tons of grain fed to livestock.

If they decided to restrict livestock feeding and use the grain instead to feed starving people in the Third World, governments would have to decide how to do it. Would they impose a meat tax to discourage consumption, or would they ration livestock products, the way meat was rationed in many countries during the Second World War? Those are the kinds of questions that we could be facing, if not next year, then the year after that, or maybe the year after that, as a result of the various forms of environmental deterioration that we're now experiencing in the world. We could see a part of the world move into a period of uncontrolled social disintegration.

If that begins to happen, says Brown, and "environmental deterioration and social disintegration begin to feed on themselves on a large enough scale, it will become very difficult to turn things around."

For those who think we'll be able to depend on science and technology and genetic engineering to solve our food problems, a word of caution comes from Norman Borlaug, the plant breeder whose high-yield wheats won him a Nobel Prize. "The most productive applications of biotechnology and molecular genetics, in the near term, appear to be in medicine, animal sciences, and microbiology. It will likely take considerably longer to develop biotechnological research techniques that will dramatically improve the production of our major crop species." The world cannot wait for those someday-soon miracles of the lab as armies of the hungry grow day by day. We are beginning to see the absolute upper limits of the ability of agricultural land to support human life: a sobering study by the Alan Shawn Feinstein World Hunger Program at Brown University in 1988 estimated that even if food were equitably distributed (and nothing diverted to livestock), the all-time record food production of 1985 could have provided a minimal *vegetarian only* diet to about six billion people, the projected world population in the year 2000. The same global harvest, allowing a diet with about *15 per-*

cent animal products, could feed some four billion people. A diet consisting of *35 percent animal products*, similar to that consumed by most North Americans and Western Europeans today, could be food to only 2.5 billion people — or fewer than half of today's population. Right now, if North Americans were to reduce their meat consumption by only 10 percent, it would free up more than 12 million tons of grain annually for human consumption — enough grain to easily feed all those facing famine in the world today.

For Lester Brown, avoiding a life-threatening food situation during the 1990s may depend on our ability to rein in runaway world population growth to bring it in line with food output — essentially to cut population growth *by half* by the year 2000. The time may have come, says Brown, for world leaders to issue a call to action. The U.N. secretary general, the Pope, the president of the World Bank, and national political leaders "should urge couples everywhere to stop at two surviving children." Because regardless of the angle from which you observe the critical state of the planet, you always arrive at the same bottom line — human population growth is out of control.

Our final testament to our belief that *nature is infinite* lies in those ever-increasing numbers of people on a finite planet. We are taking it on faith that nature will continue to provide. The evidence says otherwise. "Famine seems to be the last, the most dreadful, resource of nature. The power of population is so superior to the power of the Earth to provide sustenance . . . that premature death must in some shape or other visit the human race." Those were the words of the Rev. Thomas Robert Malthus in his essay on the principle of population written in the year 1798. And so it is coming to pass.

FOUR

G O   F O R T H   A N D
M U L T I P L Y . . .

Consider this: more people have been added to the Earth during the past 40 or 50 years than have been added since the dawn of man. We have become the most "successful organism" on the surface of the Earth, but what that really means is *Homo sapiens* is a species out of control. If you look at history you can see that human populations were always able to move on when land grew scarce and resources overutilized. Today, there is nowhere left to go. "We're up against the wall now," says E. O. Wilson. "We have too many people. We've got 5.3 billion, and that number will double in several decades, according to World Bank estimates. The world just can't take that number without our producing the kinds of effects that are now becoming apparent — the global environment changes and warming, ozone depletion, toxic pollution, and finally the destruction of the major habitats containing most of the biological wealth of the world."

Every year, the world adds 95 million people to its population. That is more than another Mexico to feed. Or another Canada every 15 weeks. "The farmers of the world try to feed that additional 95 million people with an agricultural system that is losing 24 billion tons of topsoil each year," says Paul Ehrlich. "And we're trying to do this in a system that is already staggering and facing climate change."

Ehrlich points out the way populations of fruit flies go on a breeding binge, increasing their numbers more and more rapidly until the food supply is exhausted — then they all die. He fears that humans may do much the same thing. In 1968, in his earthshaking book *The Population Bomb* in which he forecast a demographic apocalypse, he wrote: "The battle to feed all of humanity is over. In the 1970s, the world will undergo famines; hundreds of millions of people are going to starve in spite of the crash programs embarked upon now."

Ehrlich underestimated the potential of the "green revolution," which developed high-yield types of crops sustained by huge inputs of pesticides and fertilizers to keep the human show on the road. But nonetheless, as he and Anne Ehrlich say in their 1990 book, *The Population Explosion*, some 200 million people died of hunger and hunger-related disease in the 22 years since *The Population Bomb* was written.

Says Bill Rees, a resource ecologist at the University of British Columbia: "The green revolution has enabled us right until the mid-1980s to maintain a level of food production that was just slightly ahead of the rate of population growth on the planet." For a while, the "technical fix" worked. But now the pests are gaining resistance to pesticides, setting up the "pesticides treadmill." More and more chemicals have to be sprayed onto farmland to maintain crop yields. The new genetically uniform crop varieties are turning out to be more vulnerable than their low-yield predecessors, both to pests and to droughts and floods. And all the time, from Mexico to Sudan and India, more and more soil blows uselessly across the land, forming deserts or washing into the sea.

The green revolution is faltering. For the past several years there's been a drop in per capita food production globally. "We're producing about 14 percent less food per capita than we were in 1984," says Rees. What Ehrlich predicted for the 1970s could be waiting for us in the 1990s.

The fear is that agricultural technology may be buying us time at the cost of a bigger crisis when it comes. "The very technologies that we rely on to maintain the magnitude of our economic enterprise, to produce the food that we consume, are eroding and destroying the very basis of life on the planet," says Rees.

The specter of Ethiopia is abroad. The television images are haunting: an 18-year-old Eritrean, driven out of her homeland because of starvation, stands and gives birth to an emaciated infant. Do we want that quality of human life? Through the 1980s millions of Ethiopians became refugees as the soils of its once fertile highlands, drained of nutrients by too many farmers, blew away on the dry winds of drought. Such poignant images could fill our TV screens from a dozen more countries on the edge of ecological meltdown during the 1990s.

Until then, we have to make do with the bare statistics, the signposts to a terrible future. E. O. Wilson interprets them this way: "If this planet were under surveillance by biologists from another world, I think they would look us over and say, 'Here is a species in the mid-stages of self-destruction.' "

The world's population is growing by three people every second. The 1990s will see faster increases in human numbers than any decade in history. Over the next 10 years the 5.3 billion we have now will increase by almost a billion people. That's like adding another China to the planet. Dire projections come from the Washington-based Population Crisis Committee, a leading U.S. organization that monitors population growth. They predict that if present birth and death rates continue we will be struggling to accommodate almost 11 billion people on the Earth within the next 40 years. And that will continue in a disastrous upward spiral to 27 billion by the end of the next century. Most of those new mouths to feed will be born to impoverished Third World families who barely subsist now.

To put all of that into perspective, just look at the next 40 years, says Robert Ornstein. It means that it took 10 000 human lifetimes for us to grow to two billion, and in the course of one human lifetime, we're going from two billion to 10 billion. "The impact that each one of those billions has on the global environment is magnified by technologies like automobiles and the burning of coal and chemicals like CFCs and hundreds of thousands of other chemicals that are doing serious damage." Those are the concerns that grow daily with every new person added to our numbers: too many people equal too much pollution, too much destruction of the natural habitat and the Earth's life-support systems; the fear is that the rise will happen so fast and be so steep that the world's resources will quickly run out.

Demographers, the people who study the statistics of populations, say that throughout human history, population growth has come in surges, each accompanying a period of great technological change. They call these eras "the Great Climacterics." Each surge produces an increase in the world's population of up to a hundred-fold. The first known Great Climacteric occurred a million years ago, when advances in tool making made us much more efficient hunters. It allowed the world's human population to rise above one million for the first time.

The second Great Climacteric came between 10 000 and perhaps 4000 years ago, after the most recent Ice Age when humans took up farming crops and domesticating animals. After that, world population got up into the hundreds of millions. The third Great Climacteric, beginning with the Industrial Revolution in Europe, started in the early eighteenth century, when world population was about 600 million. We will have hit six billion people, a tenfold increase, by the end of the decade. The big question is how much further we will go.

Optimists see hope here. These explosions have happened before, they say. Ours will burn itself out, as the others did. Says Kingsley Davis of Stanford University: "In human evolution, zero population growth has been the rule, not the exception." The world's human population will soon reach a new plateau, say the optimists. But the reality is that the world is a little different this time around. Past climacterics have involved humans expanding onto new areas of the planet, draining marshes and irrigating deserts, for instance, but we've already taken over most of the usable land. More than 40 percent of the plant matter grown on the Earth's land surface each year is exploited in some way for human benefit. Equally important is the fact that past civilizations could destroy their environment and fail. It didn't matter to the civilizations of the Andes or Africa or Europe when the Mesopotamians poisoned their fields with salt brought down from the hills in irrigation canals. Nobody in China heard about the fall of Jericho. But this time around, the greenhouse effect or a punctured ozone layer or a new pest running through one of the world's new uniform crops could cause worldwide havoc and threaten what is now a global civilization. Like Ehrlich's fruit flies, or a locust swarm in the Middle East, we sink or swim together. China's plans to burn its vast coal stocks, releasing immense quantities of greenhouse gases, and to put a refrigerator in every Chinese home, accelerating the buildup of CFCs, threaten the entire planet. The profligacy of the United States has already put us in grave peril; if the Third World follows our example, it will finish us off.

Demographers predict a gradual slowdown in population growth during the twenty-first century, reaching some kind of equilibrium at about 10 to 14 billion. Some claim to see signs of a downturn already. Birth rates have been falling in most places since the 1960s, though death rates have fallen much faster. In 1973, the rate of population growth in the world (the difference between birth and death rates) dipped below a peak of 2 percent a year. It is now about 1.8 percent.

But rates are statistical abstractions. What matters are the extra mouths we have to feed each year. And the number of extra mouths increases annually. Standard estimates posit a peak at maybe 100 million extra people a year about 2020, and then a decline. There is plenty of baby-bearing potential yet. Half the world's population today is under age 24.

Before we begin to worry about the world's ability to feed 10 to 14 billion people, we need to wonder whether the population roller coaster will gain too much momentum to stop. The demographers, locked inside their statistical models, assume that the nations of the Third World, where 95 percent of the population growth will take place, are going to follow the same demographic path as the already-industrialized world.

Demographers point out that in general the poor breed more and the rich breed less, so a natural smoothing out of the population graphs in the coming decades will require that the current poor become rich, just as we did. But as the green revolution sputters, as the shanty towns of Mexico City, Jakarta, São Paulo, and the rest of the new "mega-cities" spread, and as tropical ecologies give way, nobody should be assuming that Bangladesh will follow the path of Britain or Europe or Canada.

The small downturn in population growth of the past 15 years already shows some signs of being a flash in the pan. Most of it happened in China, where Draconian two-child and then one-child policies in the 1970s and 1980s helped bring to heel the demographics of a nation of a billion people. But the social and political control possible in China is a double-edged sword, as events in Tiananmen Square in 1989 showed. And few nations could even contemplate implementing such policies. In any case, growth rates began to rise once more in China in the mid-1980s, as the baby boom of the 1960s began to have children. The one-child policy inaugurated in 1979 is now officially described as a failure.

India is the second most populous nation in the world. It has 850 million people, compared to China's 1.1 billion. Most demographers had predicted that it would become the world's most populous nation sometime in the early decades of the next century — until China's growth rate began to increase again. One Indian is born every 1.2 seconds; that's 72 000 a day, more than the entire population of Canada in one year. Almost one-third of the present population goes hungry. In the slums of Calcutta, they differentiate among "three-meals-a-day people," "two-meals-a-day people," and "one-meal-a-day people."

Despite four decades of effort by the central government to spread the message about family planning, including a notorious period of sometimes forced male sterilization in the mid-1970s, when men were dragged off to vasectomy clinics, the population surge shows no sign of abating. If the world goes the way of India, then we can forget about those nice smooth lines on the graphs. It could look more like one of Ehrlich's graphs of a fruit-fly population, soaring ever upward until, one day, the food runs out and it crashes.

Lester Brown of the Worldwatch Institute in Washington calls the current predicament of India and many other fast-growing nations in Southeast Asia, Africa, and Latin America the "demographic trap": "Rapid population growth is beginning to overwhelm local life-support systems, leading to ecological deterioration and declining living standards. Population growth and falling incomes are reinforcing each other." Says another population specialist, Phillips Cutright of Indiana University: "Forget plateaus"; they're just "wishful thinking." "Essentially," Brown says, "the choice is between launching a vigorous effort to reduce births, such as in China, or accepting a rise in deaths, as in Ethiopia." Ehrlich agrees. "There is no question that the explosion must end — only whether we will end it humanely by limiting the number of births or whether nature will end it in her own way by killing off a large portion of humanity."

On the face of it, there should be no problem about fitting the world up with contraception and getting population under control. We have the technology — and seemingly everybody has an interest in it. Yet in most places, most of the time, it fails. One reason is a mixture of religion and politics. During the 1980s, under pressure from the "right-to-life" lobby, President Reagan pulled the plug on all U.S. foreign-aid programs for family planning that involved, or might involve, abortion. The aid has not been reinstated. Financing for the U.N.'s Fund of Population Activities was cut off in 1986, and voluntary bodies such as the International Planned Parenthood Federation have been ostracized since 1984. Taking heart from their success with the government, the "right-to-life" lobby extended its onslaught to organizations that campaigned for the restoration of the funds. In the United States, the National Audubon Society found itself under attack. As soon as the organization began to talk about increasing funds for family planning, it was painted as pro-abortion. It was a vicious attack. There were articles in the right-to-life newsletters; a phone network was set up. Specific attacks against individual Audubon executives as being "baby killers" were launched. References were made to Audubon's preferring birds to babies. The society received letters from people across the country threatening a withdrawal of membership, with loss of money and with all kinds of boycotts. Another Washington-based environmental group, the National Wildlife Federation, has received hate mail, orchestrated by the National Right to Life Committee, for its policy of supporting family planning abroad. Says Lester Brown:

> These right-to-lifers are a group of ideologues as opposed to a group concerned with the well-being of people throughout the world. It's very difficult to put the brakes on population growth quickly if you don't have access to family-planning services backed up with abortion. There are no perfectly reliable contraceptive techniques in the world today. As a result, countries need abortion as a backup for contraceptive failure. Without it, I don't know any country that has achieved zero population growth.

The Population Crisis Committee in Washington claims that to stop the world's population short of 10 billion, "all the world's reproductive-age couples need access to reliable and affordable contraception by the year 2000." That is expected to cost $10 billion, with most of the money coming from developed countries. The committee added that average family size must decrease from four to two children. In April, the American Assembly (a forum founded by Dwight D. Eisenhower when he was president of Columbia University) brought together leading scientists, politicians, economists, and businessmen from 19 countries to tackle the environmental problem. Their final report criticized the Bush administration for its inadequate family-planning aid policy, stating that family planning should be supported and made available "to those who want it" throughout the world. The statement added "that no administration can be regarded as serious about the environment unless it is serious about global population growth."

The Catholic Church's opposition to birth control has helped encourage fast-rising populations in Latin America and Africa. In November 1988, Pope John Paul II reaffirmed the Catholic Church's ban on contraception. The occasion was the twentieth anniversary of Pope Paul's anti-birth-control encyclical, *Humanae Vitae*. Just before Christmas 1989, the Pope delivered a major statement on the environment, warning that the world's "reckless disregard" for its ecology "lays bare the depth of man's moral crisis." But he made no mention of the population explosion as an agent of ecological destruction. In May 1990, tens of thousands of Catholics filled a dusty field in Mexico and listened as Pope John Paul II preached a hard sermon against birth control. In a country that is struggling to curb its population's explosive growth rate, the Pope told the assembled faithful: "If the possibility of conceiving a child is artificially eliminated in the conjugal act, couples shut themselves off from God and oppose His will."

Worldwatch Institute research indicates there may now be more ecological refugees in the world than political refugees, and that the situation will get increasingly worse in the years ahead. As Lester Brown says, "We're in a situation where the degradation of life-support systems as a result of population pressure in a number of African countries is leading to an increase in death rates. For the Pope to go around saying you don't need to worry about population pressure, and that family planning is a no-no, is in the minds of many people a crime against humanity." Adds Paul Ehrlich: "It's very important to remember that anyone who is fighting providing people with contraception and getting family sizes down is simply fighting very hard to get millions or hundreds of millions or billions to die early, and in very nasty ways, in the not-too-distant future."

Muslim countries, notably those in Southeast Asia, include some of the most heavily populated and ecologically ravaged nations on earth. Many of them remain committed to national policies to increase their populations. Procreation is a national duty. And in other countries, fear about the growth of Muslim minorities has impeded progress to stabilize populations. Europeans, for instance, are afraid that they will soon become a minority inside the Soviet Union because of the high birth rate among the Muslims of Soviet Asia. There are similar fears in Nigeria, Africa's most populous nation. At current trends, Nigeria, like most of Africa, will have doubled its population to 200 million by 2020, and will be the world's third largest nation by 2050.

There are complex cultural issues and taboos as well. In 1988, Nigeria launched a new population policy aimed at a target of a maximum of "four children per woman." On the surface, the policy did not appear to be outrageous, since four children was the average for developing nations in the late 1980s. The plan brought a public outcry. Women's groups said it would encourage polygamy, which is legal and widely practiced in the country, especially if couples had four girls and no boy. "These modern messages do not go well with our religious and cultural traditions," said Aboyami Fajobi, executive director of the Planned Parenthood Federation of Nigeria.

The critical question is how family-planning policies can be made to work in the beds of the large and fast-growing countries such as India and Nigeria. India was the first developing country to launch a national population-control policy, in 1951. Since then it has spent billions of dollars and claims to have prevented perhaps 70 million births. But a recent study found that only 39 percent of couples use contraception. And population increase since 1951 is approaching 500 million people. The country's professed target of zero population growth by the end of the century seems a long way off.

India's government has tried every method to persuade its population to take up birth control. Sanjay Gandhi, eldest son of then prime minister Indira Gandhi, offered transistor radios to men who would accept sterilization. Publicity campaigns have used pictures of weary, underfed young mothers to focus the issue on health and have argued that small families are richer families. Now India is considering a return to a system of incentives. One model may be provided by neighboring Bangladesh, where fertility has become a traded commodity. There, during the 1980s, the government began to offer cash for sterilization. Applications soared when food became scarce during the floods of 1984. In some places, at the time, women were given food aid only if they agreed to be sterilized. Despite such efforts, Bangladesh still has among the highest birth rates in the world.

For many, these experiences show that a deeper investigation of the motives behind the determination of Third World rural families to have children is essential.

For most peasant families in the Third World, it makes sense to have more children. In the rich world, children remain an economic drain on a household almost until they leave home. In the Third World, they are swiftly converted from mouths to feed into hands for work. From the age of six or seven, they are looking after chickens or running errands. Says John Caldwell, an influential Australian demographer, people in the rural Third World don't buy washing machines or vacuum cleaners, which are expensive and impossible to maintain; they have children to do the same chores. Children also substitute for a pension plan. A World

Bank study found that between 80 and 90 percent of people surveyed in Indonesia, Turkey, South Korea, and Thailand expected their children to care for them in their old age. And they can be like a ticket in the national lottery. "Third World parents can always hope that the next child will be the one clever and bright enough to get an education and land a city job," say Frances Moore Lappé and Rachel Schurman from the Institute for Food and Development Policy in the United States. "In many countries, income from just one such job in the city can support a whole family in the countryside."

City-dwelling children are out on the streets from a young age, selling trinkets, rummaging through rubbish dumps, begging, or — notoriously, in cities such as Bangkok and São Paulo — selling their bodies as prostitutes.

However many condoms are parachuted into the Third World by aid programs, and however many billions of dollars are spent on propaganda, unless the rural poor of the Third World — the great mass of humanity — can see personal advantage in smaller families, the population explosion will continue. (And if they don't gain from it, what right does the rich world have to persuade them to make their lot even worse?) The truth is that in the Third World, having more children is a defense against all the threats to survival. Looked at this way, then, all those family-planning experts can easily become part of the problem, rather than part of the solution. As Frances Moore Lappé and Rachel Schurman concluded in their study *Taking Population Seriously*, "Rapid population growth is a moral crisis because it reflects the widespread denial of essential human rights to survival resources — land, food, jobs. . . . Far-reaching economic and political change is needed to reduce birth rates to replacement levels. Such change must enhance the power of the poorest members of society, removing their need to cope with economic insecurity by giving birth to many children."

Is this a counsel of despair? Surely the world will have descended into chaos long before we can re-arrange the planet's resources — give the poor 80 per-cent of the population the 80 percent of the resources they deserve rather than the 20 percent they are left with now. No, say Lappé and Schurman. People who feel secure about the future, and that of their family, will not need the insurance policy provided by a house-ful of children — however poor they are. We are talking power and not pesetas. In their research, Lappé and Schurman identified seven societies where population stabilization has made the most headway. Of these, four — China, Sri Lanka, Burma, and the large Indian state of Kerala — are among the world's poorest. In each case, they say, "the political, economic, and cul-tural changes that allowed population growth to slow dramatically did not depend on first achieving high per capita income."

China, despite extreme poverty during the 1970s, provided social stability on its communal farms, with every family having its own plot and with most chil-dren receiving education. This, Lappé and Schurman say, was as much responsible for the massive fall in the Chinese birth rate at that time as any more overt aspects of population control. Quite independently, John Caldwell singled out two societies on Lappé and Schurman's list — Kerala and Sri Lanka — as the two places in Asia where grass-roots organizations have pushed most effectively for "people power," with result-ing dramatic shifts in population growth.

In the southwest Indian state of Kerala, people have won the right to social-security payments, pensions, and unemployment benefits, to health care and "fair price" shops to keep down the price of rice. Land reform has got rid of rapacious landlords, even if the poor still have little land to farm. Rather than investing in expensive technology and trying to boost production as other Third World areas have done, Kerala sought to better its lot by economic redistribution.

Kerala is poor, and its population density is three times the average for India as a whole. Yet infant mortality is one-third the national average. Many Keralite women and their infants receive a free meal each day at village nurseries. In 1988 these feeding centers served 265 000 women and infants. By 1987, school feeding programs provided daily meals for more than two million children. Most children attend school. Almost three-quarters of Keralites can read or write. Less than half the population is literate in the rest of India. The Keralite female literacy rate is two and a half times the national average. The life expectancy in Kerala — 68 years —is closer to the U.S. average of 75 years than to India's 57 or the 52 years of other Third World countries. In cash terms, the people are not rich. But they are healthier, better educated, and more secure than peasants elsewhere in India. Their feeling of security shows in the smaller number of children they bring into the world. Say Lappé and Schurman, "With these few facts about life in Kerala, we can begin to understand how one of a poor country's poorest states could have achieved a population growth rate not much higher than Australia's." The birth rate at 22 per 1000 is only a little above the U.S. rate of 16; the figures are 43 for low-income countries and 32 for India.

For those many people who worry that the only way to ensure human survival is through Draconian laws of compulsory birth control and Big Brother propaganda assaults on the rural poor of our planet, here is another way. It involves liberty and justice and poor people fighting for their rights. It makes family planning a right and not a duty. And it works. The good news from

Kerala is that the population explosion is curbed when people live healthy, secure lives without the need to invest in bearing many children to assure their economic well-being. "In the 1990s," says UNICEF Director James Grant, "Third World countries ought to be meeting tangible human targets, assuring safe drinking water, assuring access to health services, assuring basic education." That is certainly the message from Kerala.

But there's another, bigger message from the population crisis we find ourselves in, and we don't have to go far afield to find it. "The most common misperception of the population problem is that it's a problem of poor Indians who don't know how to use condoms," says Paul Ehrlich. Actually "the most serious population problem in the world is right here in North America."

One North American does 20 to 100 times more damage to the planet than one person in the Third World, and one rich North American causes 1000 times more destruction, according to Ehrlich. He bases those figures on 1987 U.N. statistics on per capita energy consumption. People who drive gas-guzzling luxury cars, air condition their homes, and live on what Ehrlich calls "high-intensity-the-hell-with-tomorrow agriculture" wreak more havoc on the planet than any Third World person. "Babies from Bangladesh do not grow up to own automobiles and air conditioners or to eat grain-fed beef." The average North American uses the energy equivalent of 10 tons of coal a year. Bangladeshis use less than 100 kilograms (220 lb.). Put another way: every three-child family in North America is about as dangerous for the planet as a 103-child Bangladeshi family. The excess two million babies born each year in the United States will grow up to consume as much of the world's resources and pollute the global atmosphere as would 200 million Bangladeshi babies.

It's a message neither the average North American nor our governments seem to be hearing. The cultural mind-set in this part of the world looks at overpopulation strictly in terms of Third World multiplication. That attitude permeates Western thinking, and our lack of comprehension of Ehrlich's contention that *we* are the population problem makes one wonder whether our actions place us at the Mad Hatter's tea party while the rest of the world slides into a living hell. Consider:

— the self-indulgent parochial moralism of anti-abortion "right-to-life" lobby groups who campaign against abortion at home and family-planning aid for Third World countries, as 30 000 children a day die somewhere in the world of starvation;
— the West's infatuation with high-technology birth for infertile couples, spending as much as $25 000 to $30 000 a try and countless public funds attempting to grow test-tube babies in an already overpopulated world;
— the concern of Western governments that Western populations may be falling below replacement levels. The concern is predicated on the theory that more population means more consumption, which means more economic growth.

Both France and Germany have produced a plethora of incentives to encourage more births. The Canadian government lamented Canada's falling birth rate in a 1987 booklet *From Baby Boom to Baby Bust* in which the authors laid out scenarios for increasing baby-making in Canada and warned of negative consequences if the numbers didn't increase:

Should sub-replacement fertility persist or become more pronounced, the perception that the steady erosion of the demographic base of the nation poses a threat to its economic well-being or even its survival may gain momentum. As an alternative to fertility,

immigration in numbers far in excess of the histori-
cal levels required to maintain even a moderate population growth may be unattainable or attainable at
considerable social cost. Public opinion could then
eventually tip in favour of a more pro-natality stance,
and the call of Henripin and his colleagues for "a
plan of action" may become more acute.

We saw in Romania how extreme a "plan of action" for
baby-bearing can get. Desperate to increase Romania's
population, the Ceausescu regime made it illegal for
any woman under 45 with fewer than five children to
obtain an abortion. A secret police force, nicknamed
the "pregnancy police," administered monthly check-
ups of female workers to make sure they didn't abort
pregnancies. A special tax was levied on unmarried
people over age 25 and childless couples who could not
give a medical reason for infertility. Forced to bear children they didn't want, Romanian women simply abandoned them. As the Iron Curtain crumbled, the West
got a glimpse of the cost of the ultimate baby-making
regime — thousands of infants and children living
deprived lives in state orphanages.
    We've also seen another kind of "plan of action" in
the province of Quebec. Quebec has the lowest birth
rate in Canada. The decline is a major concern for the
province; the fear is that if the low rate continues
unabated it may presage the death of Quebec culture
and the French language in North America. So in 1988
Quebec set out to lure women into having babies by
paying financial incentives. Quebec families were
offered a $500 premium for the birth of a first child;
$1000 for the second, and $4500 for third and subse-
quent children. A year after the incentives were
announced, the Quebec government gloated that its
baby-bonus plan was paying dividends: some 90 000

babies were born in Quebec in 1989, a 6 percent increase over 1988.

In its April 1990 budget the Quebec government increased its "bucks for babies" program to a $6000 bonus for third and subsequent children.

On a planet that is already overpopulated, according to most basic ecological criteria, policies like Quebec's underscore a fundamental flaw in Western thinking. What we've all got to realize is that we're part of a larger world. "The issue is no longer can you or I afford to have more than two children, for example; the issue is whether the planet can afford for any couples to be having more than two children, beyond replacement value," points out Lester Brown. We can ask that same question in relation to every aspect of our Western lifestyle: not whether you or I can afford a large automobile but whether the planet can afford the CO_2 released by the gas-guzzlers. What is called for is for all of us to look beyond our individual and local circumstances and weigh all these issues at a global level. What we consider acceptable at an individual level cumulatively promises disaster for the planet, because, in the end, if we do not deal collectively with the population problem, it is now clear that nature will do it for us.

And if we do not quickly understand that we are one species among millions that must fit into a web of life, our sheer numbers will strip the Earth of its ability to support us.

FIVE

...AND DOMINATE
THE EARTH

---

*G*enesis 9:2 — "And the fear of you
and the dread of you shall be upon every beast of the Earth, and
upon every fowl of the air, and upon all wherewith the ground
teemeth, and upon all the fishes of the sea: into your hands are
they delivered."

*E*arth Day 1990 — And the word
went out to the people: "The Earth exists for the human person
and not vice versa . . . . The focus must be on the sacredness of the
human person . . . not on snails and whales."
— Cardinal John O'Connor
Roman Catholic Archbishop of New York
St. Patrick's Cathedral, New York

In the long history of the Earth, there have been episodes during which large numbers of plant and animal
species have been eradicated. Some of these episodes
were cataclysmic events, such as the huge asteroid that
scientists theorize slammed into the planet 65 million
years ago and put an end to the 150-million-year reign
of the dinosaur.

Now we are in the middle of an extinction spasm that
makes the demise of the dinosaurs seem minor in comparison. Humanity is carrying out the greatest massacre this world has ever witnessed. Scientists are
rarely known for their unanimity, but Harvard biologist

E. O. Wilson says there are two areas on which almost total agreement has been reached: that global warming as a result of the greenhouse effect is inevitable if we continue on our present course; and that widespread species extinction is guaranteed if we go on destroying the environment.

Nowhere is the destruction more evident than in the rain forests, those ancient lush tropical forests with their distinctive green canopies that cover only 6 percent of Earth's land mass, but harbor more than half of all species on the planet.

Within the next 24 hours, as we go about our daily lives, 54 species of animals and plants will disappear forever — at least 20 000 species a year are lost as the rain forests of the world are destroyed.

Tropical forests have existed in place for more than 100 million years. Now these primeval landmarks may disappear forever within our children's lifetime. Our children will inherit only tattered remnants of the great tapestry of life on this planet. Wherever you look — South America, Africa, and Southeast Asia — almost .60 hectare (1.5 acres) of rain forest vanishes every second. That adds up to between 16 188 000 and 20 235 000 hectares (40 and 50 million acres) a year, an area almost as large as West Germany. The methods are crude — bulldozing, clear-cutting, flooding, torching — and the reasons behind them shockingly understandable. Yet it is chilling to realize that people are converting the richest ecosystem on Earth into a pasture for cows.

With the trees go the species that make up most of the life on the planet. In some places less than one square kilometer (.386 sq. mi.) of tropical forest contains more native tree species than exist in all of North America. A mountain ridge in Peru was recently clear cut — something that can be done with chain saws in a matter of days — and when the cutting stopped, says E. O.

Wilson, 90 species of plants became extinct. For Wilson, who has spent his career studying the life of the rain forests, the horrible irony is "the loss of 90 species in the Amazon region went virtually unnoticed."

We would do more than take notice if we understood what these distant forests mean to our lives. What happens in them is critical to the future of all life on Earth. If you ask Randy Hayes of the Rain Forest Action Network, he can offer an explanation of just how "critical": you need to care about the tropical rain forests in the Amazon or in Central Africa or in Southeast Asia "if you want to breathe, if you want to eat, if you want medicines."

One-quarter of our pharmaceutical drugs come from plants whose primary habitat is the rain forest. Seventy percent of the cures for cancer now being studied involve plants that grow exclusively in tropical rain forests. In the 1960s, leukemia proved deadly for four out of five children who contracted the disease. By 1985, four out of five were surviving because of the rosy periwinkle, a plant found in the rain forests of Madagascar. A substance in this plant is pardoning children from a death sentence. The plant family that the rosy periwinkle belongs to, the genus *Catharanthus*, has six species in Madagascar. The other five are still unstudied, and one of those is on the verge of extinction because of deforestation.

Millions of heart patients owe their lives to the rain forest. Two life-saving heart drugs, digoxin and digitalis, are extracted from the foxglove. Yet fewer than 1 percent of tropical plants have been screened for medicinal properties.

The life-and-death battles fought on a daily basis in the rain forest among its millions of different species hold secrets that could help humanity. The chemical weapons used by plants and insects against their predators can assist us in waging war against disease in ways we can only begin to imagine, and may never know.

For instance, there's a moth that migrates across Panama in such enormous numbers that they blacken the sky as they fly over. These moths lay their eggs on a particular species of jungle vine: the caterpillars that hatch eat the leaves of the vine. After about three generations, the leaf counterattacks, and the caterpillar dies off.

It is the leaf's chemical arsenal against the caterpillar that intrigues Michael Robinson of the Smithsonian Institution. The compound acts as a feeding suppressant to locusts; if you sprayed it on grass, locusts probably wouldn't eat it. It also seems to imitate the chemical structure of sugars, yet it isn't a sugar, and so has potential as a sugar substitute in diet foods.

Robinson suggests that this compound could be investigated as a possible anti-viral agent, for example, against a virus like AIDS; at the molecular level, viruses hang on to sugars, and this compound might act as a decoy for the virus. Just from this one moth-plant relationship, all sorts of possibilities arise.

As important as the rain forest is as a source of pharmaceutical relief for human ills, it is the role of that ecosystem in planetary terms that cannot be overlooked. The rain forests are, in effect, "the lungs of the planet," helping to regulate the exchange of oxygen and carbon dioxide, just as our own lungs do. In the Amazon region alone, 75 billion tons of carbon are filtered out of the air by trees.

Some scientists have proposed that one of the solutions to the greenhouse effect is to plant 1.9 million square kilometers (720 000 sq. mi.) of new forest. "It's crazy to plant all that new forest," says Robinson, "if we're cutting down the old forest, which is already doing part of that job." As those trees are cut or burned, they release their carbon into the air as carbon dioxide, exacerbating the greenhouse effect.

That's something that should concern not just Brazilians but all citizens of the world, says Brazil's

Science and Technology Secretary Jose Goldemberg. Rain-forest destruction is right now contributing 7 percent of the carbon dioxide that is accumulating in the atmosphere. "You could argue that Japan contributes more than 7 percent, and Eastern Europe also, but this particular 7 percent comes from a single place where destruction could be stopped rather easily."

Computer simulations published in the journal *Nature* in 1989 raised more fears that the rapid deforestation of the Amazon would upset global and regional weather patterns. If large tracts of rain forest continue to be replaced by pasture, rainfall in the region could be dramatically reduced — the researchers predicted by as much as 20 percent.

That result would be felt worldwide. Brazil's current Environment Minister Jose Lutzenberger has gone so far as to suggest that if the forests are destroyed in the next 20 years, "the whole weather system might collapse."

The disappearance of the rain forest would be nothing less than an incalculable disaster for the whole planet if we only consider its value as a potential medicine chest, or as a climate modulator. But there is a more fundamental question to be asked of ourselves and that is whether we can allow a billion years of Earth's history to virtually disappear during our generation's lifetime. That is exactly what is at stake in tropical forests that represent most of the biological diversity on this planet. In our belief that we should dominate and control the Earth, we have lost the understanding that we are part of a web of life. "We should never forget that even as we head to Mars in spacecraft, we are of the Earth. We are organisms that evolved on Earth in a matrix of that biodiversity that we are now destroying," says E. O. Wilson. The rain forests are a vast reservoir of unique life, home to perhaps 80 000 plant species, and possibly 30 million animal species, most of them insects. Wilson discovers a new species of ant, his specialty, just about every time he does research in the rain forest. One tree in the Peruvian Amazon yielded 43 different kinds of ants. "That's approximately equal to the number of different kinds of ants found in all of the British Isles, and there they were — on *one* Amazonian tree."

It's research like Wilson's that shows how crucial every species is to the web of life. He points out that ants create soil and keep it fertile. Together with termites they make up one-third of the animal biomass — the total weight of animal life — of the Amazon forest. Closer to home, ants disperse the seeds of nearly one-third of New York's herbaceous plant species. If the 8800 ant species and their several million invertebrate relatives were suddenly to disappear, the physical structure of the forest would degrade. Most of the fish, amphibian, reptile, bird, and mammal populations would crash to extinction for lack of food. "The Earth would rot," says Wilson. "As dead vegetation piled up and dried out, other complex forms of vegetation would die off, and with them the last remnants of the vertebrates."

Terry Erwin of the Smithsonian Institution found that one hectare (2.5 acres) of Peruvian rain forest yielded 41 000 species of insects, more than a quarter of them beetles. For all our advances in genetic engineering, and given the possibility that one day we might be able to replicate life in a lab, these endangered species are irreplaceable. No one has dreamed of a technology that duplicates the immensely complex and ancient ecosystems that the tropical rain forest represents.

In a single species you would find as much genetic information, if you could translate it into English, as you find in all 16 editions of the *Encyclopaedia Britannica* published since 1768, says Wilson. "So when one plant species on a mountain ridge in Peru goes extinct, when one butterfly is allowed to go to zero population on Mauritius, that is the kind of genetic information that's lost forever."

Wilson is not talking about information for information's sake — interesting, but non-essential. You just have to look at the food on your table. Today humanity depends primarily on about 20 species of plants for food, mainly those that our palaeolithic ancestors encountered haphazardly in meso-America and in the Fertile Crescent during the early stages of agriculture, says Wilson. "In actuality as many as 35 000 species of plants exist that could serve as food plants, and some of these, even in their wild state, prior to improvement by selective breeding, are demonstrably superior to those now in use."

All this incredible diversity now hangs in the balance.

Wilson relates a recurring nightmare he has: "I'm on some fabulous tropical island like New Caledonia and my plane is about to leave. Suddenly, I realize I haven't collected any ants, so I get back in the car and I'm driving, driving. I know there's a great forest on the northern part of the island, but I can't seem to find it. It's getting late, so I start looking for some trees, but all I can find are subdivisions. The land has all been developed. I can't find a single ant."

In Peru, the rain forest is fighting a losing battle against coca growers driven by American and European demand for cocaine. Nearly 11 percent of the Amazon rain forest is located in Peru, about 760 000 square kilometers (293 460 sq. mi.). Almost 60 000 square kilometers (23 167 sq. mi.) have been lost. They are going at a rate of 3000 square kilometers (1158 sq. mi.) a year.

One-quarter of the world's primate species face extinction as Madagascar's tropical forests are destroyed. Twenty-nine different kinds of primates are found only there, mostly in the Tropical Forest Zone. Within the island borders of this country off the east coast of Africa lie more unique forms of plants and animals than in any other country in the world. It has an estimated two-thirds of the world's species of chameleons and all the world's lemurs. Ninety percent of Madagascar's tropical forest is gone. Today, driven by poverty, farmers are clearing trees and burning them to make way for pasture at such a rate that by the year 2000 almost one-third of that country's remaining forest will have vanished.

Indonesia contains 10 percent of the rain forest in the world, about 173 million hectares (427 477 140 acres). The country loses 1.2 million hectares (three million acres) a year as it tries to keep up with foreign demand for its timber. Indonesia supplies 70 percent of the world's plywood and 40 percent of its tropical hardwood.

One-third of all the rain forests on Earth sit in the Amazon basin, mostly in the country of Brazil. It is that rain forest that has seized the world's attention. In 1831, Charles Darwin wrote in his diary about the coastal rain forest he encountered when H.M.S. *Beagle* made landfall around Bahia, Brazil:

> Delight . . . is a weak term to express the feelings of a naturalist who, for the first time, has wandered by himself into a Brazilian forest. The elegance of the grasses, the beauty of the flowers, the glossy green of the foliage, but above all the general luxuriance of the vegetation filled me with admiration.

The wonder that Darwin felt, surrounded by the tropical forest, has struck the generations of scientists who followed after him. The rain forest has been described as a great green cathedral; but it is not hushed. In fact, it's a symphony — a harmony of monkey, bird, and insect sounds — that never ends.

And there are plants everywhere, says Stephen Price of the World Wildlife Fund, plants growing on the forest floor, growing at your ankles, at your knees, at your waist, at your shoulders, and over your head. You can't even see the bark of a tree anywhere because there are orchids and bromeliads and lichens growing all over them. It is so thick with life that "we don't even know what's down there. And, of course, we're losing this at a terrible rate, and we have no idea what we're losing. We haven't described the very species that are falling under the ax and are burning up in flames." Less than 10 percent of the Brazilian coastal forest that Darwin admired in 1831 remains today.

The dread possibility of a treeless Amazon is now becoming real. Altogether, an estimated 14 percent of the original Amazonian forests have been destroyed. In 1988 alone, almost 125 000 square kilometers (48 000 sq. mi.) of Amazon rain forest were burned to clear land.

And that's where we come to the other side of the rain-forest story — the endless cycle of poverty and despair locked in a deadly embrace with environmental degradation. It is the story of developing countries struggling to exist at the most basic level.

It is the story of Brazil: a country of 140 million people, most living in squalor with no land and little food; a country suffocating under a foreign debt close to U.S.$115 billion; a country with one obvious resource to be exploited — the Amazon rain forest.

Brazilians see the rain forest as their agricultural frontier and as their pot of gold at the end of the rainbow. Looked at that way, the forest itself stands as an impediment that has to be chopped for timber, mined for gold, flooded for hydroelectric power, and cleared for agricultural and pastureland.

It was the BR-364 that got Brazilian pioneers to their new western frontier. The BR-364 is a highway like no other highway in the world: a festering wound carved deep into the heart of the Amazon. Paved in 1984, financed by a World Bank loan, the 1500 kilometer (932 mi.) road ran straight into the rain forest. And with it came the timber merchants, the gold prospectors, and the cattle ranchers.

What happened next was devastating. The poor and the dispossessed — some 500 000 from the slums of Brazil's major cities searching for a better life — swarmed into the Amazon via BR-364. With the slogan of the day as their rallying cry — "Land without people, for people without land" — Brazil's poor packed up their families and their belongings and set out for the new promised land.

The rain forest couldn't withstand the onslaught. This explosion of migration was responsible for one of the most dramatic environmental events of the past decade, the deforestation of the Brazilian state of Rondonia. Bordering Bolivia in the western Amazon, this state has already lost 57 000 square kilometers (22 000 sq. mi.) of forest — 23 percent of its total area — and could be completely treeless by 1995, according to the Worldwatch Institute.

The rain forest proved to be an easy kill. The trees are shallow rooted and easily pushed over. One small man or woman can cut down a tropical rain-forest tree in an hour's time, says Randy Hayes. "You see bulldozers manufactured in the United States or Canada with gigantic chains between them — the kind that chain supertankers up to the docks — and they're dragging the chain along, just plowing down trees, and all the life that lives in them."

What they didn't cut, they burned. The smoke was so thick that airports in the region were closed for days at a time. Satellite photographs taken in 1988 showed as many as 6000 separate fires burning throughout the Amazon on a single day. Stephan Schwartzman, an anthropologist with the Environmental Defense Fund in Washington, D.C., remembers being in the Amazon during the burning season in the early 1980s when for weeks and weeks there wasn't a cloud in the sky, but you still couldn't see the sun for all the smoke. It was, for Schwartzman, a vision of hell.

There was a time when even ecologists used to think that the rain forest was very resilient, says E. O. Wilson, "a green paradise that could never die." It was thought to be so hardy that you'd have to fight against it constantly to avoid it being overgrown, but, in fact, the settlers discovered that the opposite was true.

What the would-be farmers didn't realize was that the rain forest is a very fragile system, in most places growing on extremely poor soil. They didn't understand that the trees act as sponges, retaining the nutrients. Once the trees were gone, the soil dried up and blew away. So the settlers farmed for a few years and then had to move deeper into the Amazon. Ranchers used the land for a short time to raise cattle to sell to countries with high beef consumption, then had to find greener pastures because the soil couldn't support vegetation for grazing. Ranching there was profitable only because of massive government subsidies.

Otherwise, there were no winners in the great Brazilian rain-forest land rush. Native tribes living in the jungles were dispossessed, the rain forest was decimated, and the settlers remained impoverished. What we're left with, says Stephen Price, "is the destruction of the richest, most diverse ecosystem on the planet, all for unprofitable beef." It is estimated that the cattle-ranching subsidies amounted to more than $1 billion between 1975 and 1986, making it the biggest known subsidy in history for ecological destruction, unrelieved by economic gain.

In every other attempt, as well, the economic development of the Brazilian rain forest has been a sordid tale of one disaster after another, all at the expense of this unique and crucial ecosystem. A huge pig-iron smelter in the eastern Amazon, the site of some of the largest mineral deposits in the world, is threatening to wipe out surrounding forests to feed its furnaces, much like what happened in the state of Minas Gerais where pig-iron furnaçes consumed nearly two-thirds of the state's forests.

In the forests of the Brazilian territory of Roraima, a Stone Age tribe is fighting for its life. The Yanomami, the last major isolated tribe in the Americas, number just over 9000. They are being exterminated by greed, victimized by the largest gold rush in history. Since 1985 some 45 000 miners have flooded into this remote Brazilian rain forest, lured by tales of striking it rich. They brought with them diseases such as malaria, which are killing off the native population; they separated the gold ore with mercury, which is poisoning the Amazon River, the world's largest river system; they carved more than 100 dirt airstrips out of the rain forest.

Brazil is still locked in this vicious cycle of destruction today. Brazil is a country that needs to create more than a million and a half jobs a year, points out Enio Cordeiro, a deputy in the Environmental Division of Brazil's Ministry of Environmental Affairs. "We cannot simply abandon the need to develop the entire country; we have to ensure that this number of jobs are created every year."

And, to foreign critics of rain-forest development, Cordeiro's reply is blunt: "This is not a Disney World for the pleasure of those who want to see a rain forest preserved."

So the dreams of developing the Amazon persist, and now there are new symbols of senseless destruction on the drawing board — hydroelectric dams. With a population that will grow by 60 million in the next 20 years, to 200 million people, Brazil is counting on hydroelectric power to be the answer to the anticipated enormous energy demand and has already identified 136 potential power sites, of which 70 are in the Amazon. Rain-forest expert Hilgard O'Reilly Sternberg of the University of California calls these proposed dams "the most serious threat to the ecosystem of the Amazon today."

"Balbina" conjures up images of an energy dream that turned into a nightmare. Located smack in the heart of the Amazon, Balbina was to be an ambitious mega-project, a dam that would supply hydroelectric power to the city of Manaus. Less than three years ago, the area was home to a vibrant patchwork of tropical life. Today an area half the size of Prince Edward Island, 2500 square kilometers (965 sq. mi.), lies beneath the dam's waters. As the waters rose, the animals of the rain forest retreated to hilltops and, isolated by the waters, died there — gruesome testimony to economic development gone awry. Lush rain forest has become a stark and barren scene; the barkless skeletons of dead trees thrust up through the water for hundreds of kilometers. The grotesque aspect of the Balbina dam was that engineers had overestimated its hydroelectric capacity. Generating a fraction of the expected output, barely half of what is required to meet the needs of nearby Manaus, Balbina is considered the least efficient hydroelectric scheme in the world. Silt is quickly building up, and the turbines are corroding in the acidity of the water. Unfazed by the failure, engineers are now proposing the construction of yet another dam to provide water to top up Balbina. It's engineering gone mad, and it has cost countless species of plants and animals as well as the extinction of two tribes of indigenous people.

Whether falling to torching, cutting, or damming, the rain forests are shrinking so quickly that ecologists are convinced we are the last generation that has a chance to save them. Approximately a quarter of the biological diversity existing as of the mid-1980s will vanish over the next 25 years, as the remaining forest refuges disappear, according to shocking calculations done by Peter Raven of the Missouri Botanical Gardens.

"All predictions are that within 50 years anything not locked up in a national park will be gone," explains Randy Hayes. "So if we're the last generation, then we simply have to get the job done. We have no other choice."

To start to save the rain forest, not just in Brazil, but everywhere in the world, it's important to understand who is responsible for what's happening. The charred remains of what were once rain-forest trees are a mute testament to our blindness. The forests are victims of the same human philosophy that has compromised all the life-support systems of the planet. All the values we consider normal and respectable in this world — growth, development, profit, and dominance — are killing the rain forest. Says Hilgard O'Reilly Sternberg:

> People have this idea of development, and the question is, can certain areas be developed this way? In fact I would go even further: can any area in the world be developed the way we are developing things? Third World countries are adopting the same destructive methods that North America did. After all, the United States is destroying the forest in Alaska and California; Canada is certainly doing the same thing with her forests. So this is a general philosophy of the civilization that we live in, of making a buck, and there's very little understanding of the economic as well as ecologic importance of preservation of things.

The pressure on Brazil to change its policies on the Amazon rain forest is coming from many sources. One powerful lever has been the World Bank, made up of

Western nations, which belatedly is recognizing the dire consequences of its loans to develop the Amazon. The bank says it will not fund any more roads into the rain forest and will examine all requests for money to determine the environmental impact of the projects.

The Brazilian government, responding to World Bank pressure and international outcry of all sorts — from governments to environmentalists to rock stars — in 1989 unveiled a plan called "Our Nature" to help protect the Amazon forest. One of the things it does is create an environmental-protection fund, and it invites the international community to contribute to it. Tax incentives for abolishing virgin forest have been discontinued, and migration to the Amazon region is being discouraged.

There is an even more exciting flash of inspiration for saving the world's tropical forests, and it is based on the premise that the trees are worth more standing than they are cut or burned. Brazil itself recognized this fact in an unpublished government report that stated that the country is burning Amazon forests to the value of $40 billion — more than a third of its foreign debt — every year.

That shocking waste is reinforced by a study published in *Nature* in 1989. The researchers calculated that the fruit, rubber, and selective logging in a hectare (2.5 acres) of rain forest could yield a steady annual revenue of $7000 (with the fruit and rubber providing 90 percent) compared to just $1000 from clearing all the land for timber. As cattle pasture, the same hectare is worth less than $3000, and that's not even counting the costs of weeding, fencing, and animal care. But that comes as no surprise to the natives of the forest.

Long before the white man entered the tropical forests, they were a hive of human activity. For tens of millennia, indigenous peoples lived and thrived in the Amazonian forests. More to the point, the forests thrived. "It's important to remember that the Amazon rain forest supported a population before 1500 that in all probability was larger than any population the area has supported until perhaps the present decade," points out anthropologist Stephan Schwartzman. "When you read the first travelers' accounts, they talk about very densely populated villages, very closely packed along the Amazon basin. There were many people living there in ways we can only imagine now, but there were enormous confederations of indigenous groups that had long-distance trade relations in an area with no cities, no roads, no modern technology — apparently afoot in an area of pristine tropical forest."

First the Indians, then the rubber-tappers . . . but what they had in common was the knowledge of how to use the rain forest without destroying its resources. Tribal peoples evolved systems of agriculture that allow the forest essentially to rest between periods of farming, so they can use the forest in a sustainable way, says Judith Gradwohl, who has documented the use in the book she co-authored with Russell Greenberg, *Saving the Tropical Forests*.

Tribal peoples can tell by wild plants blooming in the forest that it's time to plant the yams, and they know to plant the beans in an area that's been enriched by a certain kind of ash, and to plant the corn somewhere else. The rubber-tapper learned to extract latex from trees and learned from Indians how to live off the forest without destroying it. In the 1980s, the tappers began lobbying the government for the formation of "extractive reserves" where both rubber-tappers and Indians could continue to use the forest in a sustainable way. The murder of the outspoken rubber-tapper leader and opponent of rain-forest destruction, Chico Mendes, in December 1988 provoked international outrage and forced the Brazilian government to look seriously at the forest-dwellers' demands. Right now, there are 15 such reserves in the works throughout Brazil; they will be protected from deforestation, but open to latex tapping and nut and medicinal harvesting.

It is the standing living forest that provides for these people, and there are organizations and businesses who think the same philosophy can be exploited on an international scale, making it a win-win situation for everyone involved — Brazil, the rain forests, and future generations.

In a cramped office on the campus of Harvard University, thousands of kilometers away from the Amazon, the business of turning the living rain forest into lucrative enterprise is being hatched. In this unlikely setting, anthropologist Jason Clay plays the equally unlikely role of international entrepreneur — putting the renewable products of the rain forest together with businesses that can turn them into profit. Clay, the research director of an organization called Cultural Survival, has just come back from the Amazon, and he excitedly holds up bags of Brazil nuts. These are his greatest treasures, and his greatest success story so far. A Vermont-based specialty-ice-cream company, Ben and Jerry's, has created a "rain forest" flavor ice cream that will use 15 tons of Brazil nuts every couple of months. Rain-forest ice cream is just one of a plethora of products Clay is counting on from the forest. "On one recent trip to Brazil I brought out 44 types of wood products that can be used as essences and oils and soaps. We see coming out of the initial experiments of these products at least six to 10 massage oils." The point is that Brazil nuts and essences, which can be used as bases for anything from shampoo to aftershave to perfume, can all be taken without destroying the integrity of the rain forest.

In *Saving the Tropical Forests*, Judith Gradwohl and Russell Greenberg cite concrete examples of using the rain forest in a sustainable way. A Costa Rican project is raising iguanas, instead of cows, for meat. Basically, an iguana yields meat in smaller packages, and tastes like chicken. Iguana is a popular food in Latin America, so popular in fact that the large lizards were almost hunted to extinction. A West German biologist, Dagmar Werner, got the idea of raising iguanas in captivity and then releasing them back into the forest.

In 1987, Werner hatched 5300 iguanas at her research station; more than 4000 have since been released to the wild, now living about 150 animals to a hectare (2.5 acres). At that density, there is productive meat yield. If left to beef cattle, the same area would produce about 6 kilograms (13 lb.) of beef a year, and the land would be degraded in the process. If the same land is left as tropical forest and given over to raising iguanas, it will produce 9 kilograms (20 lb.) of meat without harming the environment.

The iguana ranchers get a double-barreled bonus, say Gradwohl and Greenberg. Iguanas live on fruit, so a farmer who joins the program gets some baby iguanas and some fruit trees, puts the trees in the ground, and lets the iguanas go. Within three years, they can start harvesting iguanas, and within five years they can start eating fruit. "So it's a way to revegetate the area and raise iguanas at the same time."

These ideas are a start, but just that, because Brazil is the first to admit that the biggest threat facing the rain forests is the country's huge foreign debt. In 1987 the U.N. Brundtland Commission on the Environment reported that the debt crisis is forcing Latin American nations to exploit their natural resources to pay creditors. The U.S. $115 billion owed makes Brazil the Third World's largest debtor, and the message to the rest of the world is clear: if the rain forests are indeed part of all our global destiny, then the debt-load problem must be first on all our agendas.

One solution that is paying off is the idea of *debt for nature*. The brainchild in 1984 of Thomas Lovejoy, the former head of the World Wildlife Fund, it involves buying up Third World debt at discounted prices against preservation of rain forests. Foreign banks, stuck with debt that they knew would never be repaid in full, were happy to accept less money, and Third World countries in debt were happy to get rid of the debt burden.

In one such swap, the World Wildlife Fund bought $1 million of Ecuadoran debt in 1988, held by Bankers Trust, at a discounted price of $354 000. In exchange, Ecuador agreed to protect some of its most biologically diverse tropical habitats, home to more than 1400 species of birds, up to 20 000 species of plants, and various endangered animal species, such as the mountain tapir and the jaguar.

Since Lovejoy's moment of inspiration, eight "debt for nature" swaps of this type have been arranged between developing nations and the World Wildlife Fund. The idea was endorsed by leaders of the Group of Seven most industrialized countries in their 1989 summit in Paris. Brazil, which at first resisted any suggestion of such a swap, seeing foreign offers as an attempt to control its rain forest, now has begun accepting foreign aid to help save the Amazon.

It's an idea that we can all participate in, first by pressuring our own governments to get involved to a greater degree, and then by putting our own money into the mix. The World Wildlife Fund in Canada has been participating in "debt for nature" swaps over the past few years, where individuals have been able to purchase and protect rain-forest land in Costa Rica. The project has since been expanded to include other rain-forest nations.

In a similar move, Anglo-French financier Sir James Goldsmith has developed a plan to save tropical forests by "renting" them from Third World nations. Goldsmith's premise is that these forests are a vital resource — left standing — for rich nations, so those countries should pay a fair price to preserve them. "We should enter into a fair-market contract whereby, in return for not cutting the tropical rain forests, the host nations are paid a rent by the remainder of the world," says Goldsmith in a pamphlet detailing the plan. The idea is similar to past policies in both the United States and Canada where farmers were paid not to harvest certain crops. "The amount paid would compensate the host nations for the cost of maintaining the forests and for the perceived loss of income from not exploiting them on an unsustainable basis."

What Goldsmith suggests is a contract between developed and developing nations. But ultimately what our species must do is negotiate a contract with nature itself — that the survival of the rain forests takes precedence over any use that we can make of them. "We must find a way to live in harmony with nature," says Jose Lutzenberger. "Politicians come along and say, 'We'll take small areas and call them extractive reserves.' I can't accept that, even for a moment. It means you accept the destruction and protect a few patches against assault on the wilderness."

Finding sustainable uses for the tropical rain forests of the world may be the way to escape the immediate dilemma of destruction, but in the end the solutions will lie not in our pocketbooks or in some economic or even sustainable values that we can attach to these crucial mainstays of life on Earth. In the end, only our attitudes can save the rain forest. The challenge for all of us will be to think beyond the practical, beyond our own human self-interest. After all, these forests are the repository of most life forms on Earth. "I believe that we ought to develop an ethic of world heritage," says E. O. Wilson.

We've come to care about the preservation of the Louvre or the great antiquities of Egypt. It's time we started caring about the world's biological heritage. It's ancient to start with, thousands of times older than our own cultural heritage. It's part of our heritage because it comprised the environment from which the human species arose.

As a scientist, I feel roughly in the position of an art curator watching the Louvre burn. The imagery I like to use in describing what we are doing, particularly with the destruction of natural habitats, allowing it to occur, especially in the tropics, is that we are burning Renaissance masterpieces in order to cook dinner.

While Third World countries with burgeoning populations struggle to provide the basics for their people, while countries like Japan suck up tropical timber for things as trivial as disposable chopsticks, and while Canadian multinationals like Brascan and Alcan continue to exploit the resources of the rain forest, all of us stand idly by without protest as the rich fabric of life on the planet vanishes more quickly than we can even begin to comprehend. Twenty percent of the world is living an orgy of mindless consumption, and the rest struggle to survive by destroying the life-support systems of the planet. Each one of us is responsible for the carnage of the rain forests as surely as if we were to take an ax or a match to the forests ourselves.

What do we do? Wring our hands, bemoan the fate of this repository of life, and live in a drabber, less varied world — if we manage to live at all? Or do we start to do something about it — now, in this moment that is all we have left? Once we lose it, this book of nature is gone forever. But the solutions are there, and in saving the rain forests we begin to save ourselves. And we start by deciding that environmental destruction is too high a price to pay for what we are being told is progress.

SIX

THAT'S THE PRICE OF PROGRESS

Mostafa Tolba, the head of the U.N. Environment Program, remembers that as a schoolboy growing up in Egypt he was shown pictures of factories in Cairo belching out thick smoke over the countryside, and proudly told by his teacher, "This is a sign of progress." Indeed, it was such an important sign that in the 1960s Egyptian banknotes carried a picture of smokestacks. "This was a symbol of development for us," says Tolba. Industrialization meant smokestacks, and smokestacks meant progress. "So, to come and tell us that smokestacks meant pollution . . . well, we welcomed pollution at that time, because there was no consideration of the aftermath."

In fall 1987, the president of a small South Pacific country took the rostrum at the General Assembly of the United Nations. Before the nations of the world, Abdul Gayoom made a poignant plea for the survival of his country. It was not the plea one usually hears in the great meeting hall of the nations of the globe. The future of the Maldives is not threatened by war; it is threatened by rising sea levels, the consequence of global warming: "As for my own country, the Maldives, a mean sea-level rise of 2 meters, (6.5 ft.) would suffice to virtually submerge the entire country of 1190 small islands, most of which barely rise over 2 meters above mean sea level. That would mean the death of a nation. We did not contribute to the impending catastrophe to our nation, and alone we cannot save ourselves."

Pollution and environmental destruction as the price to be paid for progress is a trade-off that the world has accepted since the Industrial Revolution. What we can no longer ignore is that the price of the Western world's progress has too often been paid by the developing or Third World nations. We have exported our polluting technologies to the Third World, and now we are exporting our pollution. The technologies we sent twinned their cities with our own choking industrial centers; now our carbon dioxide emissions are threatening to sink entire nations. Europe, America, and what used to be called the Soviet bloc countries generate 71 percent of the world's carbon dioxide. Sea-level rise and drought in the next five decades could drive more than 60 million people from their homes in countries such as Bangladesh, Egypt, the Maldives. Progress could become more costly than we ever imagined.

Sir Crispin Tickell, Great Britain's ambassador to the United Nations, paints a horrific picture of what that price could be:

> In virtually all countries the growing number of refugees would cast a dark and lengthening shadow. Within a country they would represent a dangerous element in what would be growing difficulties of social and economic management. Disorder, terrorism, civil war, economic breakdown could become endemic. Between countries and regions there would be still greater difficulties. Even if some people and governments wished to seal themselves off from the rest of the world, they could not do so. In no country or city can the rich fortify themselves for long against

the poor. Land frontiers can always be penetrated. Nor are short sea crossings a real barrier. Desperation could push Africans into Europe, Chinese into parts of the Soviet Union, and Indonesians into Northern Australia. Sheer numbers could swamp most efforts at control.

For the first time in history, we fortunate 20 percent of the world, we who use 80 percent of the planet's resources in pursuit of the good life, are being forced to face the fact that the Third World is part of an equation that will determine *our* survival on the Earth. The life-support systems of the planet know neither geographic nor geopolitical boundaries — whoever pollutes the air, the water, the soil will pollute it for the whole world. "One of the big problems that the First World, the rich countries, has is that it is not listening closely enough to the people in the Third World," says Paul Ehrlich. "Unless we listen to them and help them solve their problems, we're going to be in very deep trouble, because it turns out that to a large degree their problems are our problems, too."

In the past, we have used the Third World as a dumping ground for our problems. We've sent them our outdated technologies, the food and drugs we didn't want, the pesticides we banned. For example, the U.S., which accounts for one-quarter of world pesticide production, exports between 182 and 273 million kilograms (400 to 600 million pounds) of pesticides annually. The U.S. General Accounting Office (GAO) estimates that about 25 percent of this is pesticides either banned or severely restricted in the U.S. Added to that, we've shipped to the developing world the ultimate irresponsibility — the toxic wastes of our Western lifestyle. Every year, industry in the West produces 300 million tons of industrial waste. And we're offering the Third World multi-million-dollar contracts to act as our disposal site.

The going rate for the disposal of toxic waste in the rich world is up to $1000 a ton. In Africa, they will take a ton of the stuff for about $40. It is estimated that Western companies dumped more than 24 million tons of hazardous waste in West Africa alone during 1988.

Remember Koko? It is a sleepy backwater port on the River Niger in Nigeria. Andrew Lees, the toxics campaigner for Friends of the Earth U.K., remembers it well: "You approach the village of Koko through bush where quite a lot of palm nuts are being grown. It's hot. Everything looks quiet and peaceful, but then you see a rickety wire compound behind somebody's house. In that compound are vast stacks, three or four high, of 200 liter (45 gal.) oil drums. We estimated there were about 10 000 drums of chemicals."

The drums had been collected from all over Europe by a "waste broker" and sent to Koko in five shipments from Pisa, Italy, between August 1987 and May 1988. They were smuggled in by an Italian construction company on forged papers. The drums contained toxic sludge — PCBS; heavy metals, such as lead and mercury; solvents; and some radioactive wastes. "Many of the containers were bulging in the hot sun," remembers Lees. "You could actually hear a menacing hiss." The serpent was back in the garden, and we had put it there.

A spill in those temperatures, and the whole volatile dump could have ignited, billowing chemical byproducts, dioxins and furans, and permanently contaminating the whole area for miles around.

Sunday Nana, the man who had rented out his backyard for the dump site for a few dollars, said that he had heard the drums "popping" but had no idea what was in them. But the Nigerian government, which had recently complained angrily about a deal signed between Western waste firms and the government of neighboring Benin authorizing the burial of nuclear wastes there, was embarrassed and angry. It persuaded the Italian government to take back the drums. Later, Nigeria's President Babangida said, "No government, no matter what the financial inducement, has the right to mortgage the destiny of future generations of African children [by accepting toxic-waste imports]."

Koko was one incident among many. As chemists probed the contents of its hissing drums, other investigators uncovered a trail — planned and completed — of toxic-waste shipments to West Africa. Guinea-Bissau had contracted to dispose of 15 million tons over five years. Gabon had offered to take uranium tailings from the United States. The Congo had recently canceled a deal to take a million tons of solvents from an obscure company registered in Liechtenstein. Greenpeace, which has followed the toxic-waste trail, wanted an outright ban on the trade and was disappointed when an international convention signed in mid-1989 by more than 100 nations agreed only to attempt to control the trade. As the details of the convention were being worked out, two American brothers were in jail; they had amassed large volumes of waste subsequently disposed of in India, South Korea, and Nigeria. As the convention was signed, a ship carrying incinerator fly-ash from Philadelphia sailed the seas, looking for a home for its cargo.

The fear was that whatever international deals were signed, the incentives for this lethal trade would increase during the 1990s. Martin Khor, a leading Third World environmentalist and head of the Third World Network, based in Penang, Malaysia, expressed the fear this way: "The greening of the North will lead to the export of the environmental crisis to Third World countries. The trade in toxic waste from rich to poor nations has been going on for many years. It began after the United States tightened its legislation of toxic waste at the beginning of the 1980s and companies found that Third World countries would take it. The second wave came when European Community nations tightened their rules.

"You are also exporting hazardous industries. Bhopal was one example," says Khor. Mitsubishi, the electronics giant, set up a plant in Malaysia to extract yitrium from the tailings of the local tin mines. The process produces a radioactive waste called thorium hydroxide. The industry, Khor's researchers established, moved to Malaysia after its safety was questioned in Japan. "Now that the United States is banning all asbestos products, Canada is pushing its asbestos in the Third World, including Malaysia." We're foisting our hazardous wastes and industries and products on those least able to cope, just because it's convenient and cheap and there's a buck to be made.

As the century enters its final decade, commoners of the world's affluent nations live like the royalty of yesteryear, says Alan B. Durning in his report on poverty for the Worldwatch Institute:

> Yet the poor have a different tale to tell. The disparities in living standards that separate them from the rich verge on the grotesque. In 1989, the world had 157 billionaires, perhaps 2 million millionaires, and 100 million homeless. Americans spend $5 billion each year on special diets to lower their calorie consumption, while 400 million people around the world are so undernourished, their bodies and minds are deteriorating. As water from a single spring in France is bottled and shipped to the prosperous around the globe, nearly 2 billion people drink and bathe in water contaminated with deadly parasites and pathogens.

As Western parents indulge their children with the best that life can offer, the U.N. Children's Fund concludes in its 1989 annual report that "at least half a million young children have died in the past 12 months as a result of the slowing down or the reversal of progress in the developing world."

For the poor of this world, environmental conservation is just one more item in a long list of luxuries denied them. They must destroy nature to live. Land is stripped of its trees for firewood, the principal fuel used by about 1.5 billion people — more than a quarter of the world's population. The wood is cut faster than trees can regrow. Over large areas of the tropics the decline in forests is directly linked with this shortage of firewood. Developing countries find themselves growing cash crops for the developed world, while relegating their own subsistence food crops to marginal lands. The West exploits the resources at whim and at low prices. Land is cleared and devoted to export crops such as coffee and tea. But between 1980 and 1987 the prices of 33 commodities — raw materials such as copper and iron ore, and cash crops such as sugar, cotton, and coffee — fell on average by 40 percent. The free-market economy is yet one more crippling export we have dumped upon the Third World. The effect has been to make the developing world a hostage to Western debt. In 1989, the Third World owed $1.2 trillion — nearly half of its collective gross national products — to the industrial world's banks and governments. If people in Zambia had given every penny they earned to their nation's foreign financiers, beginning January 1, 1990, they would not have their debt paid until May 1993. "What people do not realize is that as we enter the 1990s, we have an international economic obscenity at large," says Stephen Lewis, the former Canadian U.N. ambassador, "and that is the flow of resources from the poorest to the richest nations of the world." In 1979, the developed world transferred $40 billion a year to the developing world. Ten years later, the developing world transferred $50 billion a year to the developed world — much of it just interest on their debt.

Lewis sees the situation as only getting worse as new international trading blocks emerge — the North American free-trade agreement, the European economic merger of 1992, Japan, South Asia, the Pacific Rim, and now the opening up of the Eastern bloc countries, — "all incestuously preoccupied with reinforcing their own wealth and privilege, and to hell with the developing world."

The industrialized nations all seem bent on following a course first suggested by human ecologist Garrett Hardin, namely, "lifeboat ethics." The idea is that a lifeboat can only hold so many people: if the Third World can't make it, well, then, we'll just have to cut it loose and let it go down the drain.

Paul Ehrlich says it will not happen that way because, as the Third World goes, so goes the entire world: "For a long time people have looked at the problems of the Third World as just the problems of the Third World, and maybe out of the kindness of our hearts we give them a hand, we give them some food, we give them some money, and so on. Now people have got to realize that they are part of a global problem; that means the thinking of the whole world has got to change. We're going to have to start working on how to prevent a large chunk of the world from going down the drain and taking all civilization with it."

What Ehrlich is saying is that the West is going to have to start paying the price of progress for the Third World. The developing world is an awakening giant, in the starting blocks of its own industrial revolution. If that revolution follows our example, it will sink us all.

And if the developing world adopts our view of progress and, as we have done, constructs a lifestyle around it, the consequences will be devastating. A single small example sums it up: the bathrooms in Khor's office in Penang are fitted out with bidets. Like most Third World peoples, Malaysians rely on water rather than toilet paper to keep themselves clean. "If 1.1 billion Chinese and almost a billion Indians decided that they need to use toilet paper, what would happen to the world's forests?" asks Khor, with blinding simplicity.

In New York, the capital of capitalism, Michael Oppenheimer of the Environmental Defense Fund brings home the moral dilemma: "We have no right in the First World to have exploited global resources and then to turn around and say that Third World countries can't have what we have. Our economic position is based on extreme exploitation of coal, forests, and other natural resources. The bottom line is that we have to all work together to find an alternative path of development for Third World countries."

The task seems gargantuan. Take chlorofluorocarbons. By the late 1980s, the world was producing some 100 million tons of them annually when the major manufacturers and their governments accepted the fact that a cut of at least 90 percent in the use of CFCs was essential to halt the thinning of the ozone layer. It had become clear to us that we could no longer saturate the atmosphere with the chemicals we unleashed from aerosol cans, refrigerators, and air-conditioning systems. Manufacturers committed themselves to finding substitute chemicals.

But where did that leave Third World nations? China had just announced a national program aimed at putting a refrigerator in every home by the end of the century. A modest enough ambition, except that with current technology it would swiftly make China the world's largest source of CFCs in the atmosphere and single-handedly undo what the Western world had achieved in trying to sew up the ozone hole. Nonetheless, China's environment commissioner, Liu Ming Pu, told a major conference on CFCs and the ozone layer, held in London in early 1989, that "CFCs are indispensable for the daily life of the masses. China needs CFCs for its economy."

At the start of the 1990s, the rich world was asking China to wait for new refrigeration technologies, to reequip its factories and invest in plants to make new ozone-friendly chemicals; progress for the Chinese would be more expensive because we, the rich world, had used up the "carrying capacity" of the planet's atmosphere for CFCs. China's response, not unnaturally, was "Sure. But you pick up the tab." India, with ambitions similar to China's, suggested the establishment of a world fund, paid for by the rich nations.

This fund would finance not only the new factories but, more important, the purchase of the patent rights to the new CFC-free technologies. It would, after all, be the bitterest of ironies if Third World nations found themselves paying patent-license fees to First World companies for the right to use technologies made necessary by the rich world's past profligate pollution.

Initially, the First World ignored the world-fund suggestion. But in June 1990, faced with new research that showed the ozone layer was deteriorating much more rapidly than previously predicted, the 56 nations that had signed the original Montreal protocol voted to establish a world fund. They pledged U.S.$240 million to help the developing world to stop producing and using chemicals that damage the Earth's ozone layer.

In very many ways, during the coming decades, the poor world will be asked by the rich world to take a difficult and more expensive route to economic development. Our own cheap road to prosperity is leading to the end of global civilization; there is no future via that route. Even if a fraction of the remaining 80 percent of the planet headed down the same highway, the ecological chaos could bring all our civilizations tumbling down. The heart of the dilemma is energy, the world's reliance on fossil fuel, and the resulting greenhouse effect. The figures are stark. The world is now producing almost six billion tons of carbon annually from burning fossil fuels, mostly coal and oil in power stations, factories, internal-combustion engines, and domestic hearths.

Right now, the United States is the major greenhouse culprit, generating about five tons of carbon per capita annually. Two billion Indians and Chinese produce less than 0.5 tons per capita; however, China and India want to industrialize their economies, taking the cheap and nasty route that brought the First World the fruits of the Industrial Revolution more than a century ago. And they can do it. Beneath their soil, these two countries have the world's largest reserves of coal. They intend to dig them up and burn them. China, which already relies on coal for three-quarters of its energy, intends to double output between 1980 and 2000, and triple it again by 2030. That would raise its output of carbon dioxide to about three billion tons annually. If India does the same, within 40 years the world's output of greenhouse gases will be more than double today's. At that level the greenhouse problem will be unsolvable. Even if we in the West were to get our CO_2 emissions under control, the Third World could push us over the brink into an age of runaway global warming. The world is effectively locked into a suicide pact.

Now that we are faced with this greenhouse apocalypse, it is extraordinary to discover that there are — researched, developed and ready to go — technologies that could clean up the First World's mess and provide the Third World with an alternative route to development.

In future decades, the world will be weaned from fossil fuels and plugged into solar, wind, and geothermal power. But, for the next 20 to 30 years, the poor nations, like the rich, can make huge strides toward a stable world by equipping themselves with energy-efficient technologies. The guru behind this development is Jose Goldemberg, a physicist, now Brazil's Science and Technology secretary and once the head of an electricity utility. In *Energy for Development*, a major report written for the World Resources Institute in Washington, Goldemberg scotched the idea that an energy-efficient world would be a world of expensive energy. His detailed studies of Brazil's plans for vast expansions in energy production led him to believe that they make neither economic nor environmental sense. Goldemberg writes that with "a total investment of $10 billion in the Brazilian electricity sector for more efficient refrigerators, street lighting, lighting in commercial buildings, and motors, it would be possible to defer the construction of $44 billion worth of new generating facilities to the year 2000." His report specifically excludes nuclear power as an option for new sources of fuel. Here, for different reasons, is another technology developed by the rich world that the poor world must be denied. To meet half the developing countries' energy needs "would require building about 100 large nuclear-power plants annually between 1985 and 2020." Apart from the cost, Goldemberg says, "such widespread use of nuclear power would entail the risks of nuclear-weapons proliferation and nuclear blackmail by terrorists because the plutonium generated in reactor operations is both a nuclear fuel and a material from which nuclear weapons can be fabricated." By 2020, such a program would produce 3500 tons of weapons-usable plutonium, enough to make perhaps 500 000 nuclear weapons. "International and national institutions would be hard-pressed to safeguard all the plutonium produced from occasional diver-

sion for nuclear weapons." Even on the brink of a greenhouse crisis, Goldemberg sees no virtue in entering the
nightmare paramilitary world of a plutonium-based
world economy.

The drawback to Goldemberg's plan is that it is complicated to organize. And that, as much as anything, is
why governments such as those of Brazil and China
have not been falling over themselves already to boost
energy efficiency. As Goldemberg's report acknowl-
edges, "Once a $2 billion nuclear power plant has been
approved, construction is a relatively straightforward
task that a relatively small, disciplined team can carry
out. But the far more complicated task of investing a
similar sum in, say, the construction and distribution of
$2 billion worth of efficient cooking stoves is likely to
require many more actors." Nonetheless, he says, with
the help of aid agencies, this route is the one that the
Third World must take. In the first sign of changing
times, in mid-1989, the World Bank, the world's largest
aid agency, threw out a request from the Brazilian government for a loan to build new power stations and
instead proposed a loan for an energy-efficiency program.

The big question for the world community is what
leverage we in the rich world should have to compel
the poor to follow a different energy path to prosperity.
During 1991 and 1992, talk will intensify on the terms
for a global climate treaty to damp down the greenhouse effect. In it, nations from the richest and most
energy indulgent, such as the United States and
Canada, to the poorest will be setting targets to limit
their output of carbon dioxide. But what kind of targets
should they be? Should there be one level for everybody? Should the rich, polluting countries be allowed
more leeway to carry on polluting than the poor world?
Should there be a global "tax" on the use of carbon, the
proceeds to be spent on investing in clean technologies

in poor countries? Should there, as some American analysts have proposed (a proposal endorsed by the Bush administration), be a worldwide distribution of pollution "credits" that could be bought and sold? So if Nigeria didn't need to pour as much carbon dioxide into the air as it was entitled to do, then it could sell some of its credits to the United States. Is this a recipe for a sane distribution of resources, or the free market gone mad?

It's the latter, says Barry Commoner, author of *Making Peace with the Planet*: "When you set up a scheme such as Mr. Bush's, and sell the right to pollute, you're setting up a market in pollutants, which means the only way the market can operate is to produce the pollutants. It is a cynical departure from the basic idea of prevention, and the same is true of a tax on pollutants. The only thing we can do is work out new systems of production that eliminate all pollutants. Don't pump them out in the first place."

Chris Flavin at the Worldwatch Institute has published his own proposals for a grand global climate convention. He wants a global carbon tax of $50 a ton. If just $5 of this were put into a global atmosphere fund, he says, that would produce $28 billion each year for investment in Goldemberg-style clean technologies in the Third World. Flavin says the two most excessive carbon producers, the United States (current annual output per capita: five tons) and Canada (current annual output per capita: 4.2 tons) should bring their emissions below three tons. Middle-range nations, such as Japan and most of Europe, should aim for a 1 percent annual reduction, and countries with emissions below 0.5 tons per capita would be allowed an increase.

The deal sounds politically plausible, but is it fair? Some say no. The targets would be easy for the rich nations to meet. A recent study found that Canada could make a net cash saving of almost $100 billion on the energy-saving measures necessary to make its first 20 percent cut in output of carbon dioxide. Only lack of political will has prevented it from happening already.

Noel Brown of the U.N. Environment Program in New York says we shouldn't forget the rich world's pollution history, the billions of tons of carbon dioxide we put into the air while we got rich. We all talk, he says, "about the Third World financial debt, but let's also talk about the 'Earth debt' that the rich countries owe to the rest of the world for past environmental misdemeanors that now threaten the planet." It might be a revealing exercise to work out how much "back tax" the rich nations owe the world.

Apart from burning coal and oil, humans contribute to the greenhouse effect every time they chop or burn down a tree without replacing it. The rich world's axmen trod the paths of the temperate forests long ago, contributing an estimated 200 billion tons of carbon to the air as the forests of Europe, North America, and Soviet Asia were felled. Today, the tropical forests are going and now the poor nations are being asked to spare the trees for the good of the planet.

As the Cold War thaws, the rich world is becoming aware once more of the huge sums of money spent in the past four decades on armaments. The talk is of a peace dividend, as the military war machines are left to rust. The swords, this time around, need to be converted not into plowshares but into heat exchangers, insulating equipment, and energy-efficient transportation. Put another way: international ecological security has replaced national military security as a global imperative. "We have no other choice but to replace a system really created to manage conflict with a system that is geared to generating cooperation on a global basis," says Jessica Tuchman-Matthews, vice-president of the World Resources Institute in Washington, D.C. Worldwide military budgets at the end of the 1980s ran at a trillion dollars a year. The Worldwatch Institute figures that we could save the planet on $150 billion a year.

The "who-pays-what-price-for-progress game" is going to have to be reformulated. The pie will have to be redivided, with the Third World getting its fair share. The first thing the rich countries are going to have to do is make helping the poor countries a major priority, says Paul Ehrlich. And that includes offering massive family-planning aid to any country that wants it; helping them get their women educated; helping to get basic medical care for children; helping to improve their agricultural systems; and changing the terms of world trade so that many of these countries can continue producing food and yet have some of the basic amenities from the industrialized world that should be available to everybody, like refrigerators. Says Ehrlich, "It's now crystal clear that we cannot solve the problems of keeping the planet habitable with our old ways of being racist, sexist, xenophobic, with great economic inequity, and so on. We've got to bring the world together so that everybody wants to work for survival."

It appears now with chilling clarity that for the first time in history, the things that used to be thought of as too utopian or too impractical are the only viable solutions to a global problem.

At the center of this call for change is the Western worldview. We can no longer afford to imagine a world in which we, in the developed countries, pursue our own economic interests and basically forget about the Third World, which we see as somehow separate and having no impact on our welfare and well-being. "What we know now is that our well-being depends integrally on what happens in the developing countries. So we will have to begin to erase the North-South line that has divided the world since the end of the Second World War and replace it with a sense of shared destiny, because we really do share a single destiny on the planet, and one that will require, for the first time, cooperation on that scale," says Tuchman-Matthews.

What we must come to understand is something very basic — that all human life on the globe is inextricably linked, says Robert Ornstein:

> People have to begin to understand that their survival and their children's survival depends on the survival of the species in a way which it never did before. There was really nothing anybody could do in East Africa three thousand years ago that would affect people in Europe or people in Asia. We need to understand that the Third World is our world, and what people do anywhere affects all of us. The Japanese appetite right now for prized hardwoods is beginning to deprive all of us of our breathing ability. In the same way the industrial countries' fossil-fuel emissions are going to heat the atmosphere for the whole world.

And the route the Third World takes to industrialization could determine the future for the whole planet.

There is a new bargain to be struck among us humans. It is called "mutually assured survival." We all share the same goal — we want the best for our children. For some that best is a very basic thing: "I would like to see my children grow well, get good education, good environment," says Mahfuzal Haque, assistant secretary in the Bangladesh Ministry of Foreign Affairs. "I don't like to see them go underwater like happened to my parents when their village home was underwater last year for 21 days. My dream and the dream of the people of Bangladesh is that at least they have the basic requirements . . . that they not drown in water. I would say forget about color television and all those beautiful things."

SEVEN

GROWTH IS PROGRESS

In a small restaurant just outside Tokyo, a sushi chef cuts a piece of tuna, rolls it up and, as the customer watches, carefully wraps a thin square leaf of pure gold around the tidbit, and presents it to the man in front of him. Picking up the sushi with his chopsticks, the customer chews the gold-wrapped tuna and breaks into a broad smile. Ah, the sweet taste of success!

Japan's is a Cinderella story of rags to riches. The nation rose from the ashes of the Second World War to become the most envied economic power on the globe. Resource poor but people rich, Japan reached into the corners of the world to pull out natural resources and, for four decades, converted them into the products we can't live without today — our TVs, VCRs, fax machines, tape recorders, cameras, stereo equipment, and on, and on, and on. In 1989, for example, a Japanese car, the Honda Accord, had become the best-selling car in America. (Long gone are the days of "see the U.S.A. in your Chevrolet.")

Life couldn't be better in Japan today. The country's economy grew at its fastest pace in 15 years in the last quarter of 1989. Along with the United States, Japan produces 40 percent of the global wealth. The Japanese have become the richest people in the world, and they are going to great lengths to find more and more exotic ways to spend their money.

Tales of Japanese conspicuous consumption are legion. Japanese travelers invade Chanel boutiques around the world in search of the classic black Chanel handbag, and then buy five or 10 at a time — at almost a thousand dollars apiece. "Japan is now the number-one market for French luxury goods in the world," says Christian Blanckaert, the president of the Comité Colbert in Paris, a trade association of 70 French companies, including Christian Dior, Louis Vuitton, Givenchy, and Chanel. Japan accounts for about 27 percent of the organization's sales. Hardly a week goes by without some objet d'art in the world going at auction to a Japanese collector, at some incredible record-setting price. With the Japanese determined to acquire objects at any price, bidding fever has left the auction world agog. In December 1989 a record was set for the sale of Chinese art, when a Tang horse went to a Japanese dealer for $5.95 million. Another Japanese bidder set a new record for Art Nouveau when he paid $1.76 million for a 1906 lotus-shaped lamp. And in May 1990, a Japanese industrialist, Ryoei Saito, the chairman of the Daishowa Paper Manufacturing Company, stunned the art world when he bid a total of $160 million for two paintings — a Renoir and a van Gogh.

It is a national spending spree that the Japanese call the *shohi bumu*, the consumption boom, and is typified by such excesses as $700 steaks or dinner for four at an exclusive Tokyo restaurant, complete with entertaining geishas, at almost $7000. Last year an American living in Japan told the story of furnishing his apartment by scavenging Tokyo's garbage. When a family there buys a new stereo or compact-disc system, the old one, last year's model perhaps, and in perfect working order, goes to the dump. Robert Craighurst, an American teacher, managed to collect a stove, an air conditioner, a sewing machine, and a color television in his forage through the city streets.

Modern Japan is a story of consumption that knows no bounds. Take gold: Japan produces almost no gold of its own, but in 1986 alone the country imported 608 tons of gold, worth $7 billion, almost half of the non-communist world's production. Shoppers are not happy to stick with conventional uses of gold — jewelry, say, or even gold coins. One company manufactures a credit card made of gold, while another produces gold house keys. Pure-gold business cards are being marketed by one Japanese firm, which sold 12 000 of them last year. Laminated in plastic, they cost $50 apiece.

The hunger for gold is so intense that Japanese are even eating it. Seiji Omura, a sushi chef, wraps raw fish in paper-thin gold foil and charges $40 apiece for the morsels. The tuna still tastes like tuna, he says, but the gold makes the dish more attractive. One Japanese restaurant is even offering a cup of coffee laced with gold dust and served from a gold kettle into gold cups at $350.

But as a headline in *The Washington Post* announced earlier this year, "Japan's wealth is creating conflict and self-doubt in the society." Many Japanese are upset by the conspicuous consumption in a culture that has long revered frugality. *Mottainai* ("it's a waste") is a word that is heard increasingly often in Japan. There is a growing feeling of disquiet that the whole society is worshiping "a golden calf."

Japan's materialistic quest has provoked questioning of its value system and goals, not only by its own citizens but by the world's. It has become apparent to all that this immense wealth and wanton spending are being financed on the back of the planet's biosphere. Japan is totally dependent on the rest of the world to supply the natural resources that support its lifestyle. But now Japan stands charged with a multitude of crimes against the natural world. A damning report by the World Wildlife Fund in 1989 made it clear that Japan holds the title as the world's worst *eco-outlaw*. "My impression is that they will do whatever it takes to secure what they need," says Campbell Plowden of the environmental group Greenpeace.

Japanese taste for gold and other minerals prompted the giant corporation Mitsubishi to offer to pay off Brazil's foreign debt in exchange for exclusive mineral rights to the gold-rich Amazon. Gold-mining in the Amazon has already pushed several Indian tribes to extinction, and threatens several others as some 80 tons of poisonous mercury are dumped each year into the area's waterways.

Japan is the largest importer of tropical timbers in the world, and it may cost the rain forests of the planet their lives. "When you're talking tropical timbers, you're talking about trees that have taken literally hundreds of years to grow," says Jeanette Hemley, director of Traffic, the wildlife trade-monitoring program of the World Wildlife Fund. "These are hardwoods, like mahoganies, and the tragedy is that the Japanese are using these tropical woods for products that are often used just once, primarily for paper production, for plywood construction forms for the building trade, and sometimes for disposable chopsticks." The rationale, of course, is the bottom line: mahogany chopsticks make "economic sense" because wood from Third World countries is cheap.

Rain-forest researcher E. O. Wilson calls it "short-term economics." "It's like saying it's cheaper to run a truck at 50 miles an hour even though you can see a few hundred yards down the line that you're about to hit a brick wall at that speed. It's cheaper right now to continue rolling up the remaining rain forests of the Amazon and Malaysia and convert them into pulp than it is to find cheaper alternatives, to continue like a giant lawnmower, just grinding up the environment."

Wilson thinks that what we have to experience world-wide is a raising of consciousness about the destructive nature of short-term economics. It may be highly profitable for a few nations, such as Japan, but very destructive to the world as a whole. The long-term costs are devastating.

The Japanese market accounts for 40 percent of woods taken from the world's jungles. At today's pace, Asian forests will be stripped of wood within 15 years. We've already experienced Japanese efficiency at clearing forests. Early in the 1960s, the Philippines was the largest supplier of tropical timbers to the Japanese market. Today the Philippines has very little forest left on any of its islands, largely as a result of Japanese exploitation.

Indonesia was next. Now Japan's focus is on Papua New Guinea, north of Australia, which has perhaps the largest continuous tract of rain forest left in the region, and on the Malaysian states of Sarawak and Sabah. In April 1990, a young Swiss, Bruno Manzer, emerged from the Malaysian jungle to carry a story to the world of Japanese destruction of the rain forest and the attempts of the Penan tribes to save their tropical home. Manzer, who had lived with the Penan for five years, told of futile efforts by the tribesmen to block logging roads. He described how in one small area of Sarawak, tens of bulldozers with two loggers each were felling about 2400 trees a day. Manzer believes that if the Japanese aren't stopped, the entire rain forest will be gone within the next five to 10 years. No forest in the world seems to be exempt from Japanese interest. Last year two Japanese pulp-and-paper companies secured 20-year logging rights to clear-cut northern Alberta forests — 15 percent of the province.

The World Wildlife Fund's indictment did not end with forests. The report charged that Japan is the world's largest importer of endangered species, both plants and animals. Rare orchids and endangered ornamental flowers and plants are imported on a large scale by the Japanese for horticultural use. The African elephant is less than two decades away from total extinction, yet until last year, Japan imported more African elephant tusks than did any other country. The ivory is bought for traditional carving and ornamental use. The craving for ivory is so strong that in 1989, after Japan bowed to international pressure to support a ban on hunting the African elephant, a British newspaper carried a story that the Japanese had turned to the Soviets for help in securing ivory — from the curled tusks of long-extinct woolly mammoths buried for centuries under the Siberian permafrost. It almost seems that extinction, or the threat of it, gives the ivory a special patina.

Jeanette Hemley continues the list of Japanese transgressions against the endangered-animal and plant world, including imports of imperiled species of crocodiles for the leather trade and rare sea turtles for the tortoise-shell trade. In fact, Japan claims 12 "reservations," or exemptions, to the international treaty on trade in endangered species. In 1987 and 1988, it imported 37 million tons of shell from the endangered hawksbill turtle — more than any other nation — mainly for jewelry.

Even with an international whaling ban in place, Japan (along with Iceland) insists on carrying out what it calls "scientific whaling." In 1988, "in the name of science" (a rationale not accepted by the rest of the world), Japan caught 241 minke whales, and killed three times that number in 1989. The minke is not on the endangered-species list, but a Greenpeace spokesman claimed that the hunts were cutting the minke numbers by half. In a report released in June 1990, the London-based Environmental Investigation Agency claimed that some species of dolphins are being slaughtered to near extinction by the international fishing industry. The study named Japan the world's biggest killer of small cetaceans, slaughtering at least 100 000 a year.

In the wilderness of Ontario, conservation officers come upon a gruesome sight as they patrol the woods: mutilated black bears — gutted, with their paws cut off. Gall bladders and the paws of bears are going to feed an insatiable Oriental market for traditional folk medicines. When ground to a powder, the bears' gall bladders fetch more than $10 000 a kilogram. Practitioners of traditional medicines — in Japan, Hong Kong, and South Korea — administer bear-gall preparations to patients suffering from liver, stomach, and intestinal disorders. Wildlife officials have no idea how many bears have been killed for these items, but they believe they have uncovered only the tip of the iceberg. In Japan and Korea, demand for the bear gall bladders has killed off most of the Asiatic black bears.

Plants, mammals, reptiles, hardwood timbers — what all these things have in common is that they are fuel for an all-consuming fire of greed for trinkets and luxuries to amuse the inhabitants of the richest nation on Earth, providing a level of consumption that reaches far beyond comfort into the absurd. "We have to keep growing," Ichiji Ishii told the authors in a 1989 interview. He is the parliamentary vice-minister of Japan's Environment agency. "I think we should be growing and growing forever. It's my personal philosophy. Quite often materials, or amount of materials available, and the degree of happiness have a very strong correlation, so I think the more we have, the better it is."

Japan's is not a unique horror story of consumption run amok. It is simply the end result of an economic system that takes no account of ecological costs or damage, that sees unfettered consumption, or growth, as progress. It is somehow fitting that economic "success" — gold — is being consumed as food. When all the plants and animals, the soil, air, and water are despoiled, that may be all we have left.

We have to look at Japan as a symbol, says Jeanette Hemley. "Japan is just one player in what is really a global issue, but a key player because of their booming economy and the philosophy in Japan to grow, grow, grow."

No aberration in the family of nations, Japan is merely the best player in the game we've devised for the human societies of this planet. If there is one thing shared by all countries in the world, rich or poor, capitalist or socialist, First World, Second, or Third, it is that economics runs our lives. Governments rise on economic promises and fall on economic performance. We live in fear of recessions and depressions, and we're all familiar, if not intimately acquainted, with notions of free trade, market forces, and tariff barriers. Newspapers are filled with reports of economic summits, mergers and black Mondays on the stock market. At the end of the year, we wait for economists to tell politicians how well off we are so that we can breathe a sigh of relief because, once again, the economy has grown and our country is in good economic health. We have been raised with the *sacred truth* that economic growth means progress.

Yale economist William Nordhaus is co-author of *Economics*, probably the best known of all economics textbooks, which is used in introductory courses in universities all over the world:

Economics' benefits are primarily that we have available to us, today, *more* in terms of quantity, quality, variety, in terms of goods and services, public health, education. . . . What economic growth has done is expand the quantity, quality, and variety of the basket of goods you can get. When my grandfather came to North America a century ago, you couldn't get anything. You couldn't buy a radio, you couldn't buy a television, you couldn't buy much in the way of shoes; there was no central heating, air conditioning; you didn't have electricity. I think the best way [to evaluate what economic growth has done for us] is to read someone's diary from 50 years ago.

Nordhaus defines economics as an anthropocentric discipline, concerned exclusively with meeting human needs and human satisfaction. That would be fine were human needs and human satisfaction met only by material goods such as shoes and televisions and radios. But if human needs and human satisfaction mean clean air, water, food, and preservation of other living things, don't look to standard economics to solve our problems. Nordhaus maintains that there's no way we can have economic activity without some level of pollution. "You can shut off pollution only by shutting off all levels of economic activity."

Similarly, if meeting human needs and human satisfaction means solving inequities between rich and poor countries, don't look to conventional economics to address the discrepancy. Nordhaus says these are not economic problems, but are "political and moral" problems. "Economics doesn't have the answer on whether we should cut our standard of living by 80 percent and send our resources abroad to poor nations or whether we should ignore the rest of the world. That has to be dealt with in the political sphere."

And if you're worried about declining natural resources, don't look to economics to lend a sympathetic ear. Nordhaus points out that economists are optimists when it comes to availability of natural resources: "People who come from the 'limits-to-growth school' would say that we have few natural resources and they are grinding down. Economists wouldn't much agree with that. We feel that natural resources are not likely to be in danger or subject to severe economic restraints for the next 100 years or so."

Such optimism may be comforting to economic theorists, but it seems incredibly shortsighted if meeting human needs means meeting the needs of the generations that come after us. After all, 100 years includes our grandchildren's lifetime.

Economics dominates our lives and the agendas of politics and business. Yet, the world economics describes seems to have little in common with the one we live in and use the discipline to explain.

Anyone reading business weeklies or the financial sections of newspapers would conclude that the world is in reasonably good shape and that long-term economic trends are promising, says Lester Brown of the Worldwatch Institute. "Yet on the environmental front, the situation could hardly be worse. Every major indicator shows a deterioration in the Earth's natural systems. Forests are shrinking, deserts expanding, croplands are losing topsoil, the ozone layer continues to thin, greenhouse gases are accumulating, the number of plant and animal species drops, and biological damage from air pollution and acid rain spreads."

Why this inherent contradiction between economic reality and environmental reality?

"The trouble is," contends Paul Ehrlich, "that economists are trained in ways that make them utterly clueless about the way the world works. Economists think that the world works by magic. Open a standard economics text, and you'll find in chapter one a diagram of the generation of Gross National Product (GNP), the final value of all goods and services produced in an economy — our measure of economic success. Yet GNP is a man-made illusion: what we see in the textbook is a 'circular' diagram that has no input at all from the real world."

The cozy world of economics is a closed, isolated system in which the real cost of doing business on this

planet, the cost to the environment in terms of depletion of non-renewable resources and pollution, has no
place.

World Bank senior economist Herman Daly holds a
minority view when it comes to economic thinking. Daly
says that continuing to study economics only in terms of
the circular-flow model is like studying animals only in
terms of their circulation system, without ever mention-
ing their digestive tract. "Yet," says Daly, "this is what the
mainline professional economic journals insist on."

Daly thinks that the real challenge for economists is
to start to deal with the real world. "The real world is
governed by such things as mass balance (matter can be
neither created nor destroyed), entropy, complex ecological relationships . . . that is the real world. Economists
have been living in an abstract world in which these
things didn't exist. They've been living in a world of the
circular flow of exchange value."

The result of this kind of thinking, says Paul Ehrlich,
is that "economists are the only major group of schol-
ars who believe in perpetual motion." They believe in
an infinity of resources. They believe in an endless ability of the natural world to swallow our waste. "They
believe in all sorts of things that are simply fairy tales."

Bill Rees, a resource ecologist at the University of
British Columbia, says we just have to look at the great
tanker spill in Alaska to see one such fairy tale. "For all
its environmental damage, for all the tragedy it created
for local people, it added several million dollars to the
U.S. Gross National Product. So it goes down by our
standard indicator of progress as a great benefit,
because it created new jobs in the shipyards that will
have to repair the tanker, hundreds of new jobs in
terms of the people cleaning up the mess, and so on.
All of those things are added to GNP, when in fact the
quality of life for people there, and indeed for the

globe as a whole, has deteriorated. Well, this is an absurd system!"

Economic indicators like GNP or GDP (Gross Domestic Product) are based on the exchange of goods and services that "create wealth." But they completely discount the environment and human activities that serve as the glue for societies around the world. GNP pays no attention to the relationship between creation of wealth and environmental degradation. If people are exposed to toxic pollutants in the workplace, or through accidental release, and suffer health problems, this is an economic benefit in that it requires the services of nurses, doctors, diagnostic labs, hospitals, and all that increases the GDP. Or if victims of environmental degradation die, this creates further work for undertakers, casketmakers, gravediggers, and so on, and adds to the GDP. Environmental degradation and activity to clean it up add equally to the economic health of society.

That kind of fairy-tale economic thinking has given us a skewed view of the world, and has pushed us to the verge of a calamitous future. We have imposed an artificial intellectual construct, a human-created design, on the Earth at the expense of physical reality, and have then ravaged the globe for economic return.

There may be up to 30 million species on this planet, of which *Homo sapiens* is but one. Yet that one species has devised a system that measures value in its own terms exclusively. Economics is the ultimate form of this conceit. If we can think of a use for something, then it has economic worth. If we can't, then it's worthless.

So in economic terms, the Amazon rain forest (or, for that matter, any forest), the bottom of an ocean, or the Antarctic continent is "undeveloped" and therefore has "economic potential"; yet, as environmental philosopher John Livingston has written, "to the millions of species that have lived in a rain forest for millions of years, the forest is already *fully occupied* and *fully developed*."

In an economics-driven world, the ecological services performed by a standing forest — cleansing the air, modulating weather and climate, preventing erosion and flooding, supporting animal and plant communities — have no meaning. In economic jargon, those services are called "externalities"; they don't figure at all in standard types of analysis that economists use to define the world. Air, water, soil, planetary biodiversity — all are considered a part of a near-infinite "global commons" shared by all and "external" to economic calculations. It's no wonder, then, that the president of a multinational forest company bluntly stated that "a tree has economic value only when it's cut down." It therefore makes economic "sense" to log a forest, bank the profit, and accumulate interest rather than to let the forest stand.

Today, our species alone has the power to affect the rest of the 30 million species on the planet. Almost overnight, we can destroy entire ecosystems; but conditioned by the long-standing resilience of nature and seduced by economic thinking that does not factor nature in to its equations, we pursue growth at the expense of survival.

We turn up evidence daily that the equation is false, as species go extinct or as human enterprise drives them to the edge of extinction. Yet economists have painfully inadequate answers to the question of how economics deals with extinction. When challenged with the premise that economics is so limited that it has no equation for the value of an entire species, the eminent American economist Kenneth Boulding replied: "I contribute $25 a year to the Save the Whales Fund. That's about what they're worth to me, and if I can convince enough people to kick in the same amount, we'll be able to save them." William Nordhaus' answer is that species extinction "is one of those externalities undervalued by the market." But while the market *undervalues* species extinction, 54 species a day disappear forever from the

Earth because of economic enterprise. "We need to do some basic rethinking," argues Herman Daly. "When vital issues like the capacity of the Earth to support life have to be classed as *externalities*, it is time to restructure basic concepts."

One of the most serious problems we have on the planet today is educating economists, says Ehrlich. "Less than 1 percent of the economists in the world have even the vaguest idea of how the world works, and yet the politicians listen to the economists."

David Pearce is one economist who believes the current economic framework can be used to save the planet, and he has the sanction of the British government to do it. Pearce has formulated a plan to ensure that the price of goods reflects the true cost of environmental assets, such as the atmosphere, the ozone layer, the soil or rivers or seas that were damaged in their use or manufacture. For all the conservation measures in the world that we can urge on people, Pearce thinks that price is "the most powerful weapon we have" to save the environment. A carbon tax that would add greatly to the price of coal and oil, for example, will result in less use of the carbon-dioxide-emitting fuels.

To ensure the survival of the African elephant, it will cost the world about $100 million a year, according to Pearce's calculations. This figure includes the cost of buying off ivory traders, paying game-park wardens, buying and maintaining vehicles for patrols, and equipping personnel properly. Pearce is certain that he can place an economic value on just about everything, but the science journal *Nature* strongly disagreed with him. It claimed that Pearce's proposition makes sense in restricted fields but that it "quickly becomes nonsense" when we have no objective basis for evaluating the damage done by pollution or other environmental insults. The journal cites as an example putting a price on the continued survival of whales or tigers. The value would probably be negative because whales eat krill, small shrimp-like creatures, which could support commercial fisheries, while tigers kill people as well as other creatures.

Nature also cites the ultimate example: "Since the major cause of environmental damage is people, should births be taxed? And on what basis?" Despite such objections, it is clear that Pearce is trying to force economics to come to grips with the real costs of our actions. What may be the serious flaw in Pearce's approach is that he thinks he can do that within the framework of conventional economics. In an article he wrote in 1988 for the British magazine *New Scientist*, Pearce said: "The only way to get the environment onto the economic agenda is to demonstrate that the environment matters to the economy. This has to be the priority for the science of environmental economics. The rest, including a more rational appreciation of environmental function and values, will follow." That statement is hollow when you realize that the notion of integrating the environment into the economy is backwards. The economy is really a subset of the natural world. So it is the economy that has to fit into the environment, not the other way around.

The idea that the economy has an independent life is a cultural myth, one of our most cherished illusions, says Bill Rees. "We made it up." Economics developed at a time when human population was small and the world was considered an infinite resource, its air, water, and soil an infinite sink for our wastes. It was an easy assumption for the new science to make because the human economy barely made a dent in the natural world.

Economics can trace its roots to the sixteenth and seventeenth centuries when Descartes, the father of modern science, gave us the concept of the division of mind and matter. This established the idea of the scientist as an objective observer of an external reality. "Well, we've carried that right over into our economic system," claims Rees, "so that economics has always regarded the environment and the resources it represents as a separate external thing that provides free goods. . . . It never considered that we depend on that natural world to actually survive."

Since 1950, the world economy has quadrupled in size. Right now, human activity alone is consuming 40 percent of the natural productivity of the land. We've got a situation, points out Rees, where a single species among the millions on the Earth, through the growth of the economy, has grabbed almost half the biological productive capacity of the land, and it can only do that at the expense of other organisms. The "good life" for some humans has been subsidized at the expense of other life on the planet, including that of future generations.

What we have then is an economic system, once a small component of the ecosystem, growing steadily within a limited space. In a world in which everything is finite, physical limitations alone dictate that nothing can grow forever, that all things will end. And where does it end? "It seems to me," says Rees, "that there's no genius involved in saying that a system growing within a larger system that is fixed ultimately results in a total collapse of the two."

That warning was first issued almost two decades ago in two controversial books that outraged economists, industrial leaders, and politicians by attacking the theological foundation of economics — the idea of growth.

The argument made in MIT professor Jay Forrester's *World Dynamics* and also in the Club of Rome's *Limits to Growth* was that an end is coming: "If the present growth trends in world population, industrialization, pollution, food production, and resource depletion continue unchanged," the Club concluded, "the limits to growth on this planet will be reached some time within the next 100 years." Nothing has happened in the past 19 years to change that prediction, says Jay Forrester. The world has continued down the same road it was on in 1971 when he wrote his book. "I think we should look upon this process very much like a biological cancer. Biological cancer grows until it kills the host on which it is living and thereby kills itself. I think mankind in the world environment is quite capable of following the same scenario." The message is that our assaults against the environment and our insatiable appetite for more of the planet's resources have caught up with us. Our attitude seems to be that "there's lots more where that came from." We've become cogs in the wheel of the huge machine of global economics whose only purpose is to create wealth, and at the expense of every other living thing on the planet, including three-quarters of humanity.

Regardless of what ideology the world has subscribed to — from Marxist materialism to capitalist consumerism — the thesis has been the same: nature is a free commodity that must be exploited and used to satisfy human needs, with little concern for the consequences. In the past several months we've witnessed how severe the consequences can be as the rusty Iron Curtain crumbled to expose an environmental nightmare.

Nowhere is the appalling impact on man and nature of Soviet-style economics more evident than in the story of cotton. It is the saga of an economic five-year plan, an imperative to grow at any cost, leaving in its wake death, disease, and devastation. It began when Moscow's economic planners decreed that it was the "internationalist duty" of the Republic of Uzbekistan to produce cotton, a valuable export crop for the Soviet Union. The Soviets call it "white gold," but for the people living around the Aral Sea, it has become "white death." In order to meet the targets for the cotton harvest (67 percent of Soviet cotton is grown in this area), the two main rivers that empty into the Aral were diverted for irrigation. Once the world's fourth largest inland body of water, the Aral Sea, deprived of life-giving waters, has shrunk by half in the past 30 years. Soviet scientists predict that the Aral "will vanish by the year 2010." The last fish died in the sea in 1983, and the excessive use of water for irrigation has left behind 26 000 square kilometers (10 000 sq. mi.) of salty man-made desert. Eight or nine times a year, dust storms drop five million tons of salt, sand, and dust on surrounding cropland, destroying harvests. A *samizdat* (clandestine document) sent to Paris described the scene in almost biblical terms: "The sky is covered by a salty curtain, the sun becomes crimson and disappears in the salt dust. In that province not one tree grows on the land. The livestock are perishing. The people are getting sick and dying."

To cultivate to the limit, massive amounts of chemical fertilizers, pesticides, and defoliants were poured on the cotton crop each year — about 600 kilograms a hectare (1323 lb. for 2.5 acres), compared with the average Soviet agricultural use of 30 kilos (66 lb.). A defoliant, butifos, that had been banned in 1983 was still being used on the crops in 1987, to "use up supplies." And all these chemicals leached back into the rivers that supplied the drinking water to the Uzbeks. Now two-thirds of the people in the Karakalpak region bordering the Aral suffer from hepatitis, typhoid, or throat cancer. Eighty-three percent of the children have serious illnesses. Infant mortality in this part of Soviet Central Asia is up to four times the national Soviet average. More than 35 million people live in the affected area. Progress for them has become a cloud of death, all in the name of economic growth.

Yet growth is what the whole world has come to live for. It is the holy grail of our political and economic systems. The measure of any government's success has become the extent to which its economy has grown. Growth and progress have become interchangeable terms. To economists, growth is the main reason governments, industries, and societies exist. Herman Daly, co-author of *For the Common Good*, calls this phenomenon "growthmania." Economic growth is held to be the cure for poverty, unemployment, debt repayment, inflation, balance-of-payment deficits, Third World poverty, pollution, resource depletion, the population explosion, crime, divorce, and drug addiction. In short, says Daly, economic growth is both the world's panacea and its *summum bonum*. A society that says, "Enough! We've got enough. In fact, we've got more than enough, so let's just coast," is simply inconceivable. But all the signs are there that it is time now to start conceiving of the inconceivable.

"The way you've got to get economists and businesspeople on board," declares Paul Ehrlich, "is for them to understand that if they don't get off this economic-growth kick, if they don't start changing their behavior, their kids, everyone's kids, are going to be dead. The ecological system supports the economic system. If we don't maintain the ecological system, there won't be any economic system — there won't be any businessmen, and there won't be any economists."

You don't have to be a math whiz or a Ph.D. in economics to see that we've reached the limits. If you go back to the figure of human consumption now gobbling up 40 percent of the natural world's resources for our own needs, and if you accept World Bank projections that human population will double within the next 40 years, it doesn't leave much room on the planet for any other than human life to operate. And as the maverick economist Herman Daly points out, "Since humans can't survive without the services of ecosystems, which are made up of other species, we're getting close to an ecological impossibility. . . . [Unless we] awaken to the existence and nearness of these limits, the greenhouse effect, ozone-layer depletion, and acid rain will just be a preview of disasters to come, not in the vague distant future, but in the next generation."

Daly's economic views are considered heretical by most of his colleagues. Many economists argue that through science and technology we can continue to stay on top of our problems, that the human mind has an infinite potential to solve the kinds of problems we have here on Earth.

One economist who believes in the human ability to overcome any problem we encounter and thus continue to grow and prosper forever is Julian Simon, whose views held sway with the Reagan administration. Simon believes that unchecked population growth is good. He complains that ecological critiques perpetuate a myth of scarcity and dwindling resources. Simon clearly states the faith held by most economists: "There is no reason why human resourcefulness and enterprise cannot forever continue to respond to impending shortages and existing problems with new expedients that, after an adjustment period, leave us better off than before the problem arose."

Simon is outspoken in his optimism: according to him our air and water have been getting cleaner rather than dirtier — the "supposed scares don't exist." When confronted with population numbers that will reach six billion by the end of this decade and almost 11 billion within 40 years, he answers, "Isn't that wonderful!" Is he worried about the numbers? "No," he says, "we're doing better every year. We're getting richer and healthier, living longer, having higher living standards, and it's all because of more people." Simon's position is that we should all stand up and cheer that "we are able to support five billion people healthily, well sheltered, well fed, where only 100 years ago we could only keep a billion alive." In his view, overpopulation is "the most incredible triumph of humanity." According to Simon, we have more resources now, not fewer, and the reason for that is the human brain — what the economist calls "the ultimate resource." "Cheer up," he says. "The world is getting better."

Herman Daly finds this kind of thinking extremely dangerous. "It's part of a general point of view that wants to say that human beings are so special that we're exempt from all the laws of nature. We rewrite them to our convenience." Daly points out that most of the basic laws of science are statements of impossibility: it's impossible to travel faster than the speed of light; it's impossible to have perpetual motion; it's impossible to have spontaneous generation of life; and so on. That's what separates science from black magic — the recognition that some things are not possible. Daly's fundamental point is if the Earth's physical resources are finite, civilization cannot be sustained indefinitely — no matter how ingenious and efficient technology becomes — unless growth stops and a steady state is achieved. That means growth in population and growth in material goods. The only way the human economy could continue to grow, points out Daly mischievously, would be "if the diameter of the Earth began to grow at a rate equal to the rate of interest." So "this notion propounded by Simon and others that all things are possible through human ingenuity is quite against the basic grain of science."

Yet Simon just carries standard economics "to some of its logical conclusions," says Daly. "He has not been rejected by the mainstream profession at all." That Simon's economic views are shared becomes clear when you talk to Milton Friedman, a Nobel laureate in economics. "We're not running out of resources. That's obvious. We have more resources today than we had 2000 years ago. You didn't have any gasoline to run your car 2000 years ago," says Friedman sarcastically. "So the simple Club of Rome approach is a bunch of nonsense because it was done by engineers, not economists."

This is the economic thinking that governments have bought hook, line, and sinker. Economists' sense of optimism comes from faith in human ingenuity and technology, but history tells another story. "Our problems do not seem to have disappeared, and we've had technology for 200 years," points out Jay Forrester. "These problems are continuously outrunning the technology."

Perhaps there's no better example of technology succumbing to problems than the so-called green revolution in agriculture that occurred after the Second World War. Miracle chemicals resulted in an incredible 2.6-fold increase in grain production between 1950 and 1984, from 624 million tons to 1645 million tons, thereby raising per capita consumption by almost 40 percent. Because cereals provide half of human calorie intake worldwide, those numbers had a significant impact on our perceived ability to feed the world.

Those four decades of steady growth of world grain output became one of the most predictable of economic trends since the Second World War. The muscle power of the human brain — science and technology — was going to enable us to conquer hunger.

In fact, the very success of the green revolution in those decades had economists worried. They talked about a world floating in grain, a "grain glut," and the notion caused concern. At the end of 1986, the World Commodity Outlook of the London-based Economist Intelligence Unit complained that the world was producing too much food, and that the developed world must deal "with the problems of existing plenty without wrecking the world trading system and thereby plunging the world into protracted recession." So much for economic prediction.

In the past few years, food growth has stopped abruptly and in fact is now falling yearly. Output per capita has fallen 14 percent over the past five years. Why hasn't food production increased during this five-year period, a time in which the world's farmers invested billions of dollars to expand output and in which fertilizer use increased by 18 million tons, more than 14 percent? "Technological advances like hybrid corn, high-yield wheat, chemical fertilizers, and irrigation — all of which boosted world crop production between 1950 to 1984 — have largely run their course," says Lester Brown. An added factor is the increasing scarcity of new cropland and fresh water that affects most of the world. Those damaging trends, in combination with the droughts of the past few years, have sounded the death knell of the green revolution.

From mid-century on, the increasing use of fertilizer has been the engine powering the growth in world food output. Fertilizer use tripled in the United States over the past quarter century, but its effectiveness has diminished. Twenty years ago, the application of each additional ton of fertilizer in the U.S. corn belt added 15 to 20 tons to the world grain harvest; today it may add only five to 10 tons.

In the United States, pesticide use in agriculture nearly tripled between 1965 and 1985. About 70 percent of all agricultural land is treated with pesticides. But despite the heavy spraying, farmers still lose more than 30 percent of their crops to weeds and pests, a rate of loss directly comparable to that experienced before the dawn of the "age of pesticides."

The heady economic predictions of continual increase of the food supply with the bonanza years of the green revolution have given way to increased human suffering instead. The U.N. Food and Agriculture Organization projects that if present farming practices continue, 64 countries — 29 of them in Africa — will be unable to meet their populations' food needs by the turn of the century, even if all arable land is farmed. Contrary to what economists such as Julian Simon claim, for millions of people right now hunger is a life-and-death battle. In this one year, more than 14 600 000 people will have lost the fight.

World population continues to grow wildly, and Jay Forrester concludes that it is inescapable that we are coming to a terrible end. "We're forestalling that evil day through ever more technology, which produces even greater problems." Fertilizers contaminate groundwater, and add to the greenhouse effect by the release of nitrous oxide. Pesticides are responsible for between 400 000 and two million poisonings worldwide every year. Up to 40 000 of them result in death, mostly among farmers in developing countries. In the United States, routine agricultural practices have contaminated groundwater with more than 50 different pesticides in at least 30 states. And all of us worry about what pesticides we're going to find next in our fruits and vegetables. The green revolution miracle cure of insecticides, herbicides, and fertilizers was supposed to solve the problem of hunger, says Forrester. "Now they have become the problem themselves."

There is a long list of other examples of human ingenuity and technological solutions that instead of solving our problems gave rise to a whole new set of problems for us — DDT, chlorofluorcarbons, nuclear power, fossil fuels, and disposable plastics, to name just a few.

"No scientists believe in the *infinite potential of the human mind*," asserts Paul Ehrlich. "The Club of Earth, which is made up of scientists who belong to both the National Academy of Sciences and the American Academy of Arts and Sciences, the two most distinguished scientific groups in America, unanimously reject the concept. And so does everyone who examines the situation. But economists, as I've said before, live in a fairy-tale world and they believe fairy tales. Scientists know better." Yet the whole world has bought in to the promise of "growth."

Japan epitomizes both the success and the destructiveness of economic growth, but we all subscribe to the same philosophy and play by the same rules. The game is the same; only the language has changed. In the 1990s our political leaders are caught in a contradictory tangle of green rhetoric and economic "realities." They pay lip service to environmental issues, but not at any cost to economic growth.

U.S. President George Bush, calling himself "the environmental president," cites the environment as an issue that has to be addressed at the highest levels of government. But then he warns a U.N. panel on the environment against costly economic solutions to global warming and policies that would interfere with economic growth and the free market. In a 1989 speech to a meeting of the Intergovernmental Panel on Climate Change, Bush said, "Wherever possible, we believe that market mechanisms should be applied — and our policies must be consistent with economic growth and free-market principles in all countries." The White House Chief of Staff, John Sununu, admitted at the time that he had altered Bush's speech to read that way because "faceless bureaucrats on the environmental side" tried to "create a policy in this country that cuts off our use of coal, oil, and natural gas."

In November 1989, British Prime Minister Margaret Thatcher made a profoundly moving speech to the United Nations in New York in which she warned that "it is life itself that we must battle to preserve. The evidence is there. The damage is being done." Then she added, we must have "continued economic growth to generate the wealth to pay for the protection to the environment." She called for the world to commit to action on climate change, then committed Britain to more and wider highways over the next decade in a white paper titled "Roads to Prosperity." Considering that the number of vehicles in Britain has risen by five million in the past decade to 23 million, and a further rise of 142 percent is expected by the year 2025, and that one-fifth of all greenhouse-causing carbon dioxide comes from cars, Mrs. Thatcher's call for commitment has a certain Alice in Wonderland quality.

But nowhere is the contrast between economic thinking and green rhetoric more startling than in Canada. Playing on the world stage, Prime Minister Brian Mulroney introduced the 1988 international conference on global warming with the words, "Among the serious problems facing industrial society, none is more acute than the deterioration of our environment." In the same speech, Mulroney went on to say that "we believe that there are no limits to economic growth, other than those imposed by our imagination, but we do recognize that there are real limits to natural systems and resources." It is impossible to understand how the basic contradiction implicit in that statement can be rationalized unless, of course, economics is clearly disconnected from "natural systems" in Mr. Mulroney's mind. Presumably, that is the way he reconciles the fact that his government will proceed with fossil-fuel energy mega-projects while acknowledging that carbon dioxide emissions will lead to global warming.

In September 1989, Mulroney opened the World Energy Conference held in Montreal by telling the assembly that the industrialized countries have a responsibility to take a leadership role in assuring the environmentally safe use of energy. "Common sense tells us that if energy demand were to be met by fossil fuels and forest depletion, it would not be environmentally sustainable," said Mulroney to delegates from 88 countries. That was for the world stage, but at home Mulroney and his ministers were practicing something else.

In the length of time it took to do a single radio interview with the authors in 1989, then Environment minister Lucien Bouchard laid out two diametrically opposing positions, both presented with the same fervent conviction. First, the minister told us that the most serious environmental problem we're facing right now internationally is global warming, and went on to say: "If we don't move now there will be a disaster. I don't want to scare people but we're dealing with the survival of the species. It is a question of great, great emergency. We must stop saying the burning of fossil fuels is the only way to live in Canada." But when queried as to why the federal government was going ahead with mega-oil projects such as Hibernia off the coast of Newfoundland, Bouchard said that Hibernia and other mega-projects had to proceed because, "while we're concerned about the environment, we're also deeply concerned with other aspects of life, like jobs" — and besides, he added, they couldn't stop projects like Hibernia now because the decision to proceed has already been approved.

Then there was the case of Brian Mulroney's Energy Minister Jake Epp. A Canadian-government-sponsored conference on the greenhouse effect recommended in 1988 that carbon dioxide emissions be reduced by 20 percent by the year 2005. But a year later, Canada's provincial Energy ministers won a reprieve from the target from Epp, who told them that, "in view of the social and other economic considerations that have to put into the mix," it would be premature to commit to the proposed cutback target. Epp voiced this escape clause almost in the same breath as he stated that the issue of global warming "requires the most urgent and concentrated attention both by government and the public." It is telling that in a Green report card on the world's seven largest industrial nations, Canada, along with Japan, tied for second last place for environmental responsibility. Only Italy got lower marks. The G-7's environmental records were scored in July 1990 by an international coalition of environmental groups. Canada flunked "because the prime minister and his colleagues are fond of green rhetoric on the international scene, but don't tend to follow through at home."

Like creatures who act according to in-built biological disposition, we who have always boasted of intelligence and foresight seem incapable of straying from past assumptions and strategies. That trait is a dangerous one that is most obvious in politicians who measure the future of the world by the length of their political term.

Bush, Thatcher, and Mulroney and his ministers all mouth the same earnest green rhetoric; all play for time, gambling with our future and the future of all life on the planet according to the rules of political expediency while maintaining the illusion of economic progress. All this goes on while they search for an *out* to the problems of environmental degradation that will be compatible with a philosophy of economic growth. And all of them have seized on a word that they believe will ultimately permit us to have our cake and eat it too. The word is *sustainable,* and for many politicians and economists and businessmen it is synonymous with *salvation.* However, some environmental analysts warn that the current interpretation of sustainable development will ensure our end. "I think the idea of sustainable development is a deception," says Jay Forrester. "It is never very clearly defined. If by 'sustainable development' we mean the continued growth of industry and population, then it is the same old process. The very word 'development' does not convey the idea of coming to an equilibrium balance with the envi-

ronment." The system is just using sustainable development to justify the status quo, says Bill Rees. "It's not
sustainable environment; it's become sustainable development, and more than that, in the minds of government and business and industry, the idea has become
sustainable *economic* development."

In 1987 the U.N. Brundtland Commission on the
Environment defined sustainable development as meeting the needs and aspirations of present generations
without compromising the ability of future generations
to meet their needs. The commission equated sustainable development with "more rapid economic growth
in both industrial and developing countries," seeing
that as the way to reduce the vast gulf between rich and
poor, and to bring the developing nations up to a
decent standard of living. Yet there's no indication that
economic growth will do the trick. The poorest half of
the world's population have been increasing their
incomes by about $7 annually, while the richest one-
fifth have been increasing theirs by about $270 annual-
ly. Ted Trainer, an Australian educator, points out that
this dismal record has been achieved "during the most
favorable period that has occurred in the entire history
of capitalism, the 1950–70 long boom." Aside from the
crucial point that we've had economic growth for the
past 200 years and it has benefited only the wealthy
West, Bill Rees believes that we're just deluding ourselves if we continue to define sustainable development
as sustainable economic growth. It just isn't in the
"ecological cards." We are suffering under the grand
illusion that we live in an infinite universe and that the
economy can expand infinitely. Rees sees the real question as being whether the planet can sustain another
round of economic growth. "It would require a five- to
tenfold increase in the nature of material, industrial
activity to bring the Third World more or less up to
European standards of living, and that's not on." More

to the point: in order to give the Third World the kind of growth it will need to survive, Rees says that it may be necessary in the next couple of decades for the industrial world not only to stop growing but "to embark on a system of rational self-limitation." Any positive growth the developing countries have will have to be counterbalanced by a reduction in the wealthier economies. Herman Daly calls it a redistribution "away from the consumption and resources dedicated to the satisfaction of relatively unimportant wants in rich countries" (such "wants" as electric toothbrushes, moving sidewalks, and color TVs) "toward the basic human necessities in the poor countries" (such "necessities" as food, clean water, and health care).

So, says Ted Trainer, the goal of development, for rich and for poor, must be to create conserver societies; that basically means the rich must live more simply so the poor may simply live.

What these economic heretics are talking about is a fundamental change in our value system. We've become blinded by the idea of progress defined by technological domination and economic growth. The challenge is to abandon the ethic of material growth as the be-all and end-all, and to embrace a new definition of growth — perhaps one in which GNP might be measured in human potential that consists of cooperation, altruism, and those environmentally sound activities that have become lost in our unending quest for material possession and new wealth.

What we've got right now is a corporate culture that places on the highest pedestal the mercantile standard by which everything is judged. "If you can't turn everything into money, which makes the rich richer and the corporations' cash registers ring, then it's not adequately valued," says consumer advocate Ralph Nader, whose answer is to create a civic culture instead — one that is a sensitive membrane for a whole range of values other than just mercantile ones, not just GNP, and the personal average income, and corporate profits. Nader lists such values as health, safety, the impact on posterity, the preservation of other living things, taking a longer view, developing a qualitative measurement of our quality of life.

If you look at history you see that growth has been part of our culture for only the past 200 years. Before that, material growth was negligible and not really important. We have been corporatized more than any other society in the history of the world, according to Nader. In ancient China, Mandarin China, they understood that merchants were important, but, says Nader, they put them on the lowest rung and told them: "Stay by yourself, perform what you have to do, but we're not going to honor you because we have other values in the society that we don't want contaminated by the mercantile class."

"We've allowed all boundaries around our merchant class or the corporate class to be demolished," says Nader, "as this tidal wave of mercantilism sweeps forward and says, 'Oh, pollution, look at it, that's the price of progress. That's the smell of economic activity,' instead of 'That's the scourge of cancer.'"

What is demanded of us is a rethinking at every level of society. We must quickly realize that to continue to subscribe to economic thinking that places humanity outside the natural world is to compromise our children's prospects; to believe that we can continue to grow materially is to push us to the brink; and to continue to conduct our lives as if nature is there for our exclusive use, *at our disposal*, is to send us blindly into oblivion.

What is being asked of us in the name of survival is nothing less than to take a new measure of what it means to be human, to face the future stripped of the cloak of materialism.

EIGHT

THERE AT
OUR DISPOSAL

A s people are confronted with the enormity of the global eco-crisis, a common response is "What difference can I make when my contribution to it is so trivial?" or "What can I do?" The magnitude of the problem is difficult to comprehend because nothing in our evolutionary history has prepared us to think globally. The state of the Earth may seem irrelevant to us compared to local concerns and the need for jobs or economic growth, or it may seem so immense that we feel helpless to do anything about it. The major challenge for each of us is to understand that the environmental crisis is the expression of the sum total of human activity. As the saying goes, if you're not part of the solution, you're part of the problem.

The call to action is muffled by the cozy assumptions of Western civilization. Occasionally, when the reality of environmental disaster creeps up to the door and knocks, we respond. A housewife in Love Canal, New York, heard the message loud and clear when she discovered that her house was sitting on top of a toxic-waste disposal dump, that local children were developing peculiar tumors, and that miscarriages among her pregnant neighbors were more common than they should have been. Lois Gibbs' courageous battle to protect her community has long since been co-opted as a symbol of personal triumph against huge establishment odds in the short chronicle of emerging twentieth-century social awareness. In 1990, merely admiring such courage is not enough. Each of us must have it. Not many of us are willing to picket utilities or lie down in front of bulldozers that we are told are clearing the path to progress. Nonetheless, the enthusiasm that greeted the Blue Box recycling campaign in Ontario and the popularity of books that tell us how to live a "greener" life have shown that lots of us are willing to do something. But the industry that is growing up around the notion of "saving the planet" is an indication that we are too ready to accept that *something* is *enough* — too eager to be lulled into a false sense of security. There is no quick fix for the environmental crisis; however, each of us has the power to make a crucial difference to the mounting odds against survival. When you ask yourself "What can I do?" the only acceptable answer is "a lot more than I realize."

And the starting point is right at home — to rethink the last *sacred truth* we live our lives by: that nature is there at our disposal.

North Americans are the most wasteful people on the face of the Earth. In Rome, people put out a little over 680 grams (1.5 lb.) of trash a day; in Nigeria, it's about 450 grams (1 lb.). In North America, every day, each person throws out almost 1800 grams (4 lb.) of waste. Over the course of a year, that's almost a ton of garbage a person.

A typical North American goes through and discards 7 kilograms (16 lb.) of junk mail and 54 kilograms (120 lb.) of newsprint each year. Each day, North Americans use hundreds of thousands of plastic tampon holders. Each hour, we throw away more than 2.5 million non-returnable, non-recyclable plastic bottles.

You don't need a slide rule or a calculator to work out that with numbers like those, coupled with rising population, garbage is changing the face of the planet. There are significantly more of us daily, and the volume of garbage we produce is astronomical. It's already up 80 percent since 1960, and it's estimated that by the year 2000, our garbage production will increase by 20 percent.

No place on Earth that has been touched by humans remains unscarred by the contact. We have left our refuse everywhere. In 1990, as part of an Earth Day celebration, a group of Chinese, American, and Russian mountaineers scaled Mount Everest to clean up some of the two tons of garbage left behind by climbing expeditions over the years. Oxygen bottles and tents were just some of the rubbish the team either burned or brought back for disposal. Organizers of the high-level cleanup crew said they wanted to leave no trace of humans on the mountain.

A 1989 marine conference held in Halifax estimated that five million plastic containers are thrown into the world's oceans every day — each of those days rued, in April 1989, by residents of a small Nova Scotia village. They awoke one morning to find on the beaches a wall 1 meter (1 yd.) wide and 8 kilometers (5 mi.) long, composed of almost every type of plastic product known and disposed of by humankind, from washbasins to oil containers, disposable lighters, and children's toys.

Until we came along, the ecosystem had developed ways to deal with the garbage generated by its inhabitants. Dung beetles rushed to collect feces, suppertime leftovers were cleaned up by birds and small mammals, death catered dinner for maggots and flies. And, in fact, for a large part of our history, the ecosystem could deal with human trash too. Now the garbage that we produce at an unprecedented rate is spilling over the limits of the planet's finite capacity to absorb it.

We have created a world where time is money, convenience profitable. We put our babies in disposable diapers, enough of them each year to stretch to the moon and back seven times. Worldwide, one billion trees annually are cut down for those fluffy liners in disposables. Every hour of work saved by a disposable diaper translates into hundreds of years of waste. Modern life is a garbage maker's perpetual-motion machine: every year 1.6 billion pens, two billion razors and blades, and 246.9 million scrap tires are discarded; and every three months, Americans throw away enough aluminum to rebuild the entire U.S. commercial airline fleet. In the United States, more than half the paper and glass produced and about one-third of the plastics are incorporated in items with a lifespan of under one year.

Breaking the garbage habit means destroying the cultural myths we have created as a society — myths that have allowed us to define ourselves by our possessions, that have led us to canonize time and convenience. The basic tenets of the consumer society are that nature is infinite and exists to serve us, and that constant demand and, therefore, constant growth are good. To keep the post–Second World War economy buoyant, an entire value system was contrived to "stimulate consumption." Disposability is profitability; items must be designed specifically to be used only once or a few times and then thrown away. Marketing consultant Victor Lebow confirmed the importance of "forced consumption" in the *New York Journal of Retailing* in the mid-1950s: "Our enormously productive economy demands that we make consumption a way of life, that we convert the buying and use of goods into rituals, that we seek our spiritual satisfactions in consumption.... We need things consumed, burned up, worn out, replaced and discarded at an ever-growing rate." It was a strategy that the fashion industry seized on early. Says Stuart Ewen, chairman of the communications department at Hunter College in

New York: "If you were to do a study of the history of product obsolescence, the model would be fashion. The first industry that built in disposability as a feature of economic survival was the ready-made clothing industry." That business was founded on the principle of premeditated waste: useful clothes can no longer be worn because they're out of style.

As we consumed, we recognized no limits: no limits to the raw materials we used to make our consumer goods; no limits to the aluminum from bauxite we used to make our pop cans; no limits to the trees we cut to make our paper plates; no limits to the energy we used to create them. "It was disposability that gave us the sense we had absolute mastery of nature," says Graham Decarie, chairman of the history department at Concordia University in Montreal. "We could take out whatever we liked. We could throw away whatever we liked. The Earth would supply us with all we needed forever. We could dump our garbage back into the rivers, and the lakes, and into the ocean, into the air, and into continents that we didn't have to look at, and it would never affect us."

But it has affected us, and that age of blindness is gone. We can't toss out our responsibility with the garbage any longer. We can't put it at the curb and forget it. The land, the air, and the water we've coopted as a dump have no more capacity for such use. Space for landfill is rapidly running out. In Ontario alone, three-fifths of a hectare (1.5 acres) of land is being irrevocably consumed by garbage each day. Forty percent of Canada's municipal dumps will be filled to capacity within the next 10 years. By 1993, Toronto will have nowhere to put its trash. The two dumpsites now in use will be full. Four states — Florida, Massachusetts, New Hampshire, and New Jersey — will close virtually all currently active landfills within the next 10 years.

What we haven't buried, we've burned; but incineration simply exchanges one kind of garbage for another

— air pollution. A "state of the art" incinerator today, burning 2250 tons of trash a day, will emit five tons of lead annually, 17 tons of mercury, 2248 tons of nitrogen oxides, 853 tons of sulfur dioxide — the list goes on. In the United States, some 155 incinerators are now in operation and 29 more are under construction, but an additional 64 have been blocked, canceled, or delayed because of such toxic-emission concerns.

We've lived the myth of disposal; now we're embracing a new one — recycling — in an attempt to recapture some of the lost energy and resources, and cut down on the amount of garbage going to landfills. In the world of garbage management, recycling is one of the three R's — recycle, reuse, and reduce. Glass bottles, aluminum beverage cans, and newspapers can be ground up and reprocessed and then recycled, at a real energy saving.

For example, making an aluminum pop can from a recycled one uses only one-twentieth of the energy required to produce it from raw materials; by doubling aluminum recovery rates worldwide, over a million tons of air pollutants, including toxic fluoride, can be eliminated. Making paper from recycled stock requires only 30 percent of the energy needed to make the paper from trees. Recovering just one Sunday's print run of *The New York Times* would leave 75 000 trees standing. Japan recycles 50 percent of its waste; West Germany, 15 percent and Canada only 4 percent. The Environmental Protection Agency in the United States has set a goal of having at least 25 percent of the nation's municipal waste recycled by 1992. A pilot study in New York State has shown that it is possible to recycle 84 percent of household waste. The motivation is certainly there. One recent poll showed that three out of four people recycle newspapers, bottles, or cans.

Those of us with recycling programs in our neighborhoods can conscientiously put our newspapers, glass containers, plastic pop bottles, and aluminum cans out for pickup and think we're doing our part. But recycling doesn't come close to dealing with the root of our garbage problem — our runaway consumption and our penchant for using things once and then tossing them out.

Richard Gilbert sits on the Canadian Federal Task Force on Packaging, and he believes that recycling programs are misleading, that recycling can't be equated with good garbage management: "Recycling is just a sophisticated twist in the throwaway society. It's number three on anybody's waste-management list. It's deceiving us. It's saying you can still continue to consume as much as before, you just have to throw the stuff into your garbage can in a slightly different way. And it's not hitting at the real issue — our obscene level of consumption. Recycling is just reinforcing the throwaway society."

Gilbert can see a society where *reuse* becomes an important way of life, a strategy that can move us toward a no-waste society, "one where very little is used and everything that is used is reused, and reused and reused, and then it's unpicked and put back together again and reused. That way, when you have something, you make it last as long as possible."

If recycling glass bottles to be ground up and remade uses 25 percent less energy than creating a bottle from virgin material, imagine the energy saving, not to mention the pollution that isn't being created or the resources that aren't being used up, if those same glass bottles are simply washed out and reused for pop or milk. Standardizing glass containers for food would go a long way toward creating that kind of saving. For example, a 500 milliliter (17 oz.) salad-dressing bottle could be exactly the same as a 500 milliliter ketchup bottle. If all our glass containers were standardized, we'd be conserving fuel and at the same time decreas-

ing truck emissions: bottles would not need to be
returned to one company's outlets.

Reuse is heading in the right direction, but the journey is futile without restraint: the most efficient way to
avert the looming garbage crisis is to avoid producing
garbage at all. That's called reduction, and as far as
Richard Gilbert is concerned, the best place to start is
with packaging. "Half of the waste we put out is packaging; out of our homes, out of our businesses, out of
our factories. We use twice as much packaging a person in North America as they do in Europe, and most
European countries now have a higher standard of living than we do. That's where the real inroads into this
problem are going to be made." In 1986, Americans
spent more for packaging than farmers received in
income. Producing these packaging materials con-
sumes 3 percent of the national energy budget.

Is it really necessary that every dozen screws you buy at
the hardware store be packaged in a square of plastic and
cardboard? There was a time when you simply bought
them out of a bulk box and put them in your pocket. A
society is grotesquely extravagant indeed that produces
and pays for a huge amount of material for which there is
no real use beyond advertising and marketing.

Starting with Berkeley, California, in 1987, a number
of municipalities have joined the rush to ban plastics
and packaging. The Berkeley council banned all
polystyrene, and more than 100 state, county, and city
jurisdictions, including New York and Los Angeles, have
followed their lead. In Minneapolis, they've banned all
plastic wrappings and containers for both prepared
foods and those sold in retail outlets. Nebraska has
banned disposable diapers. Maine has passed a law banning the use of single-serving juice boxes (more than a
billion individual foil-lined drink boxes, complete with
shrink wrapping and a plastic-encased straw on the
side, are sold in the United States each year).

The average Canadian family goes through about one ton of packaging a year, and about 80 percent of that ends up in landfills or incinerators. Provincial environment ministers have endorsed a program to cut packaging by 20 percent by 1992, 35 percent by 1996, and 50 percent by December 31, 2000.

As the movement to ban plastics grows, plastics companies have whipped their chemists into a research frenzy to find ways to make their product acceptable. They have gone in two directions, one biodegradable, the other recyclable. They're searching for the technofix for garbage, but that approach doesn't factor in resources or energy used to make the plastics, or the pollution created in its manufacture — pollution, by the way, that is toxic. In an EPA ranking of the 20 chemicals whose production generates the most hazardous waste, five of the top six are those commonly used in the manufacture of plastics for packaging.

So if we can't bury it and we can't burn it, and recycling only reinforces the myth that we can continue to consume as much as ever as long as we throw it out differently, then for survival's sake, we must add a fourth R to the garbage-management primer — *rethink* — and that, in the end, is what will save us.

What we've got to rethink is our consumer society, says Richard Gilbert:

We're just using too much. We've got to cut down; we've got to move toward the no-waste society. We've got to understand we cannot continue to burden the planet with the refuse of our lives. Garbage has become a metaphor for consumption. It's an expression of our affluence, and it has been throughout history. The real challenge before us in our society is to separate that until now almost indissoluble link between affluence and waste. We know we're rich because we can afford to throw things away.

We've got to do something that's never been done before in history: we must find a way to live waste-free. "It's a challenge that really is not addressable by technology; it's only addressable in our own minds," Gilbert says.

That challenge goes far beyond the problem of garbage accumulation. In a consumer society garbage is only an outward symbol of a fundamental skewing of values, the belief that nature is at our disposal strictly for human use; we pay for the trinkets of our lifestyle with capital borrowed from the planet, and the loan is being called in.

The subtexts of the consumer value system — materialism and planned obsolescence — join forces to create need. The latest fashions; the latest gadgets; the latest "look" in cars, furniture, or beer — daily we're bombarded with advertising that cajoles and seduces us into keeping up. We believe that our possessions make a statement about who we are. We scan ads and catalogs and magazines for an identity, "expressing" ourselves in pickled-pine cabinets, Reebok sneakers, and Rolex watches. "The whole of the advertising sector is dedicated to turning people into consumers. We don't read about 'people' anymore in the newspaper," says resource ecologist Bill Rees. "A bumper sticker that you see around Vancouver — HE WHO DIES WITH THE MOST TOYS WINS — is a symbolic representation of the games people play in our economy right now." A study conducted in 1989 as part of 3SC Monitor, a Canadian research program that measures social change in Canada and abroad, indicates that the same attitude will prevail among young people through the 1990s. Young people "wear" products as "badges" of the social identity they wish to project. What they are doing is "using consumption to define themselves," the study said. But young people are no different from the rest of society. Age is not a factor in the worship of the Gucci god.

Western society has made keeping up with the Joneses a cultural *raison d'être*. It's a way of life that puts tremendous social value on having, rather than thinking or doing.

But the sale is over; the racks are emptied, and tomorrow, when the doors open, there will have to be a new attitude on display — one that emphasizes conservation and thriftiness.

"You know, I can see fashion 20 years from now, maybe 10 years from now, not being something that you wear just for the afternoon or just for the summer. You'll be in style if you can say that your jacket will last for 100 years," says Richard Gilbert.

Clothing that stays in style for the same length of time that green garbage bags last in landfill is not the only lifestyle change we'll have to make. Our garbage comprises the total effluent on which our industrial society is founded — the one million tons of toxic chemicals released annually in the United States (among them, 60 cancer-causing agents); the toxic waste that we try to foist on Third World countries; the greenhouse gases that are the exhaling breath of our industrial society. These are the bags we leave at the curb of the natural world. "Why is it that our generation in the 1980s and 1990s has the right to reach back through millions of years of geologic time to get deposits that fuel our civilization, and then quickly transform them into pollution that will be here for thousands and hundreds of thousands of years into the future?" asks U.S. Senator Al Gore. "Don't we need to think about those who come after us?" That's really the bottom line, isn't it? How do we get up in the morning and look our children in the eye and tell them that we spent their future?

Will it be possible to make the fundamental change? Thomas Berry, head of the Riverdale Center for

Religious Studies in New York City, has "deep forebodings" about our capacity to change.

> We have a great opportunity to do significant things, but there is the question about how we're going to keep it authentic, how we re going to act on the proper order of magnitude, because people think that by recycling, that by better waste-disposal processes and so on, we can improve the present industrial pattern of our economy and so get by with it. But that is a total illusion, to my mind, because the present system is simply too devastating. The pattern has to change, our priorities have to change, our professions have to change, our education has to change and there's little indication that anything is being done.

Bill Rees believes we're in need of a new worldview, and that it will require a "massive effort at every level of society to change the value set, the expectations of people by which we now operate on a daily basis, if we're going to move in the direction necessary to save the species, and save the planet."

Moving in that direction means reevaluating the totems of our lifestyle — consumerism and convenience — and declaring them to be taboos. It means a fundamental shift to an iconoclastic worldview where human identity is not an adornment or extension of the self, but a part of the larger integrity of planetary interconnectedness. It means a focus away from the disposable to the enduring. It means understanding that from the minute we get up in the morning till we go to bed at night, everything we use and everything we discard has an impact on the planet. In concrete terms, it means we must ask ourselves hard questions about what we can afford on a planet of finite resources. Perhaps the hardest question of all that we must face is whether, given the state of the world, there can be a future for the automobile.

Among the household gods of self-definition, the car is supreme. It is our statement to the world about who we are. We have a love affair with it, but that love affair could turn out to be a fatal attraction. The number-one environmental problem is parked just outside our door — the consummate symbol of our convenience culture.

Through its burning of fossil fuels, the car contributes to the greenhouse effect and to acid rain; through its air-conditioning systems, to the ozone hole; through its plastic bumpers, plastic consoles, rubber tires, metal and composite bodies, to the garbage crisis; and through its planned obsolescence of a new model every year, to our "throwaway" mentality.

More than 400 million vehicles clog the world's streets today, and the production and use of fuels for automobiles accounts for an estimated 17 percent of all carbon dioxide released from fossil fuels. In one year, the average North American car gives off its own weight in carbon; as much carbon as it would take 20 hectares (49 acres) of forest cover to absorb — and all that is adding to global warming.

"If there is anything in our society that is the enemy of the environment, it is the automobile," says Richard Gilbert. "The most important thing that anyone can do about the environment right now is to sell his or her automobile."

It's a big step toward averting environmental collapse, but taking it is a daunting prospect.

In 1916 James Doolittle described the car in his book *The Romance of the Automobile Industry* as "elegant in lines, powerful in action, wide in service." The modern automobile, he said, "represents the incarnation of the transportation art — the silent, always-ready servant that has more strength than Aladdin's genie, and that already has accomplished vaster works for mankind's betterment than anything that has gone before."

But the car has become more than a technological marvel designed to ease the burden of daily life; more than a mechanical servant able to do your every bidding, to put you in touch with your family, your doctor, and the movie theater. The car has been elevated to the level of fantasy.

You only have to look at car ads to see how our egos have become one with the automobile. We invoke the wild and the untamable in what we name cars — Mustang, Barracuda, Jaguar — and fantasize about the thrill of power and control that riding in an enormous sheath of metal gives us. You can trace the transformation of the car in American culture through films coming out of Hollywood, says Julian Smith of the University of Florida. The car "went from being a mode of transportation, in the ordinary sense of the word, of moving you bodily through space, to being a mode of emotional 'transport,' or ecstasy. As Hollywood and Detroit came of age, they both learned how to supply dream vehicles that would carry us away from danger or boredom, transport us to better themes and bigger adventures."

The car has become a symbol of freedom, a symbol of independence, a symbol of self-reliance, says Laurence Goldstein, of the University of Michigan and editor of the book *The Automobile and American Culture*. That symbol has been exported all over the world. The world automobile population has been growing faster than the human population. It has increased from 50 million vehicles in 1950 to the more than 400 million today, almost a tenfold increase in 40 years. Projections of global numbers place almost one billion cars on the planet by the year 2030. You would think that at least in the United States, which already has a much greater per capita vehicle population than anywhere in the world, the automobile population would start to level off, but in fact it continues to increase.

These ominous automobile numbers will explode as the Third World rushes to catch up with the "American Dream" machine. Imagine one billion Chinese deciding they all want to have a car. It could happen. Governments in many developing countries feel that an auto-centered transportation system is essential to growth. They see the car as indispensable as a spur to economic growth and as a cornerstone of industrial development. In many of those countries, national resources are marshaled in their entirety to build and maintain an auto-based transportation system to serve a privileged minority of the population.

China and India together account for 38 percent of the world's population, and at this point, they own scarcely 0.5 percent of its automobiles. That situation is likely to change, given their governments' priorities. Until the late 1970s, cars were among the lowest development priorities in China and India, but since then governments in both countries have formulated policies geared toward emulating the motorized transport systems of the industrial West. The number of cars in China has risen tenfold, to half a million, and is likely to keep pace as wealth distribution becomes less egalitarian.

As yet, though, Third World car pollution is a minor-league threat compared to what we've achieved in that quarter in the industrialized West. Right now commuter traffic in New York burns more energy in one week than all of Africa uses in a year. Or, looked at another way: the entire fuel consumption of the Third World, for all purposes, is only slightly greater than what the First World uses for its cars, trucks, and buses.

It's the thoughtless squandering of energy to fuel our cars and our blind denial of the pollution we create when we drive them that epitomize our shameless lack of concern for the environment. Every time we hop in the car for a Sunday drive, jump in the Jeep for a quick trip to the corner store, or join the rush-hour caravan to work, we deny the limits of our atmosphere and thumb our noses at common sense.

Even with emission controls, the amount of garbage we spew into the air with our cars is incredible. Every motor fuel available today pollutes. Gasoline is one of the dirtiest. Cars that run on gasoline emit unburned hydrocarbons and oxides of nitrogen that react with sunlight to create low-level ozone, the major component of smog. Automobiles account for about 30 percent of total CO_2 emissions in North America and for virtually all carbon monoxide emissions in urban areas.

You might think all the environmentally damaging emissions spill out of your car's tailpipe. But it's not so. Car air conditioners, some 60 million in the United States alone, are the largest single source of chlorofluorocarbons in the atmosphere. CFCs, besides destroying the ozone layer, are now recognized as one of the most potent greenhouse gases. They contribute almost 20 percent of the total global warming effect; that's a larger share than what is attributable to the burning of gasoline.

Cars are also major contributors to acid rain. Motor vehicles emit 47 percent of the total nitrogen oxide responsible for acid rain. It is estimated that in Canada alone, with current car-exhaust standards, acid pollution from tailpipes will go up nearly 25 percent by 2005.

Then there's the oil. If you were to pay the real cost of oil at the pump, you'd have to factor in the environmental degradation that comes with oil exploration — the cost to wildlife, fisheries, tourism, not to mention the cost of oil spills. The Environmental Protection Agency estimates that there are more than 10 000 oil spills each year. In 1987, there were 15 005 reported spills. During the period 1973 to 1984, 660 403 040 liters (174 460 567 gal.) of oil were spilled in U.S. waters. The Coast Guard estimates it costs $10 a gallon to clean up oil spills. This would mean the government cleanup cost during this 10-year period was almost $2 billion.

According to a study done for Environment Canada, a large oil spill can be expected to hit in or near Canadian waters every one or two years.

And who is to blame for the oil spills? Considering that most of the oil is intended for use in our vehicles — cars, trucks, buses, and airplanes consume 63 percent of the oil used in the United States — one Greenpeace ad pointed the finger quite accurately. Under the picture of the *Exxon Valdez*'s captain, a caption read: "It wasn't his driving that caused the oil spill, it was yours."

"We have to encourage our governments to stop lying to our people about prices," says Irving Mintzer, energy-policy analyst at the University of Maryland:

In the United States, we've been lying to people systematically about the cost and the appropriate price of gasoline. Today, in my village of Washington, the average price of gasoline is lower in real-dollar terms adjusted for inflation than at any time since before the Korean War — even though we all know that neither the scarcity value of gasoline nor the total environmental impact of gasoline use has declined during that period. As a result, in the past few years, we've seen in the United States, in Western Europe, in Canada, and in a number of other countries an increase in total energy demand as people respond quite rationally to historically low prices. In North America this has meant people buying cars that have lower efficiency than they have had in the past, making the second-car choice in their family a pickup truck or a four-wheel-drive vehicle. I have many friends in Washington who never drive over any terrain rougher than a parking lot but own a four-wheel-drive all-terrain vehicle that can drive up the side of a mountain.

Let's face it: in North America we're used to getting where we want to go, when we want to go, and how we want to go. We like the convenience of making any number of stops along our own route and not having to rely on someone else's schedule. We like the privacy of our own vehicle. It's hard to give up that freedom. In fact, a study done for the Toronto Transit Commission (TTC) revealed that even if public transit were easy to use, cheap, and efficient, people wouldn't give up their cars. The survey found that most of those who didn't use public transit were men aged 25 to 54 with an annual income of $50 000 or more. Over half of them said they loved their cars so much that even if fares were free they still wouldn't ride the TTC.

Many car companies already have prototype cars that get between 80 and 100 miles per gallon (34 and 42 kilometers per liter). Analysts for the American Council for an Energy-Efficient Economy have recently proposed an efficiency goal of 45 mpg (19 kpl) for cars, and 35 mpg (14 kpl) for light trucks. If those goals were achieved, 1.2 million barrels of oil a day would be saved.

But with the number of cars projected to increase around the world, attempting to decrease emissions of greenhouse gases, carbon dioxide, and carbon monoxide, as well as nitrogen oxide by fuel efficiency alone is impossible. This year Americans will drive 25 billion more miles than they did last year. Not only are they the biggest travelers on the planet, but they travel interplanetary distances. In 1990 the U.S. auto and truck fleet will travel two trillion miles — the distance to the planet Pluto and back — 364 times. And some expect miles traveled to increase by about 3 percent a year overall, and by as much as 6 percent in many of the most polluted urban areas. At that rate, the total number of miles driven could nearly double in 20 years.

For that reason, alternative fuels are being touted as the big solution to the pollution problem. The one favored by the Bush administration is methanol, an alcohol fuel made from coal, wood, natural gas, or garbage. But depending on the gasoline/methanol mix, even that alternative fuel will contribute to global warming. None of the proposed alternative fuels — for example, natural gas, ethanol (an alcohol fuel made from corn and sugar cane), or electricity — is without pollution or technological problems to overcome.

The real savior of the North American car culture may be something that's still a long way off: hydrogen fuel. It has two basic benefits: it is environmentally benign, and it is plentiful. When hydrogen burns, it produces almost nothing but a puff of steam. The fuel can be made from water, using electricity from solar cells, and can be stored for future use. Mercedes has built a prototype that can run on solar energy and hydrogen, but a car that can travel any distance has yet to be developed.

Meanwhile, fuel efficiency and alternative fuels are an empty dream of a society desperately denying the ecological impact of its way of life.

Cars are an environmental disaster well beyond what comes out the tailpipe or the air conditioner. Michael Renner has been looking at the role of the automobile and its effect on the environment for the Worldwatch Institute:

Large stretches of land have been taken over by the automobile and its infrastructure. Parking a car at home, the office, and the shopping mall requires on average 372 square meters (4000 sq. ft.) of asphalt. More than 155 388 square kilometers (60 000 sq. mi.) of land in the United States have been paved over: that works out to about 2 percent of the total surface area, and to 10 percent of all arable land. Worldwide, at least one-third of an average city's land is devoted to roads, parking lots, and other automobile-related elements. In American cities, close to half of all the urban space goes to accommodate the automobile; in Los Angeles, the figure reaches two-thirds.

Places like New York and London, which already devote 25 percent of their land to roads, are nearing their capacity to handle more cars. Traffic in London moves as fast today as it did in the days of the horse and carriage.

Not that it really makes much difference. Policy makers are beginning to realize that adding more roads does not solve the problem of congestion. In fact, it just attracts more cars. A 1988 Department of Transportation report in California concluded that even spending a whopping $61 billion on a road-building program couldn't solve the state's traffic problem — nor in fact could any kind of road-building program.

As the car overruns us, more and more people are becoming convinced that the true solution is much more brutal than fuel efficiency and alternative fuels. We must restrict our driving. Metro Toronto Councillor Jack Layton feels we must start restricting car use or our lungs won't survive. He insists, "by the middle of the next century, the automobile has to be a thing of the past. To get there from where we are now is the challenge."

Already some cities have taken specific measures against the automobile. Drivers who want to enter downtown Singapore during rush hour have to buy a $30-a-month sticker, or carry four or more passengers. In Hong Kong, cars are equipped with electronic sensors that price highway travel by time of day. Commuting hours are the most expensive. Drivers are billed at the end of each month. In the center of Rome and Florence, all extraneous traffic has been banned between 7:30 A.M. and 7:30 P.M.; the exceptions are buses, taxis, delivery vehicles, and cars belonging to those who live in the area. Mexico City, Athens, Amsterdam, Barcelona, Munich, and Budapest are also restricting automobile movement.

Stockholm may become the first European capital to charge for road use, with the money collected devoted to improving public transport. Manhattan has now forbidden construction of new buildings with parking garages. Toronto City Council has singled out the car as a place to start in a campaign to make the polluter pay. Toronto is the first city in Canada to tackle global warming by reducing the production of greenhouse gases. The city has committed itself to achieving a 20 percent reduction of the 1988 levels of carbon gas emissions into the atmosphere by 2005. And to do that, everything connected with cars is going to become more expensive, including parking, fuel, and licensing fees.

Taxing and increasing the costs of the automobile in general will decrease car use somewhat, but as the next generation takes to the road any gains made by such policies will disappear. As the International Energy Agency, part of the OECD in Paris, points out: "Suppose a substantial tax on the carbon content of fuels were introduced, increasing consumer prices by at least 20 percent. This would slow carbon dioxide emissions in Western countries. But even with a 20 percent price hike, emissions would still be 13 percent higher in 2005, instead of 25 percent higher."

Now, Los Angeles, the city that spawned a thousand megatrends in lifestyle and entertainment, is setting the most critical trend of all — stating bluntly and without compromise that nature is no longer there at our disposal. Every month in the L.A. basin, they have an oil spill as large as, or larger than, the *Exxon Valdez*. In other words, every month, except for a very brief winter hiatus, roughly 378 540 000 liters (10 million gal.) of petroleum-type products are put into the air in L.A., causing the city to fail federal clean-air standards every two out of three days. "That's why kids in L.A. have a 10 to 15 percent decrease in lung capacity by the time they leave high school," says Larry Berg, one of the originators of L.A.'s 20-year plan to make the city breathable. "On some smoggy days, it's the same as breathing battery acid."

Los Angeles is now saying flat out that it will no longer accept pollution as the price we must pay for having convenience in our lives. This city of 12 million is shouting to the world, WE'RE NOT GOING TO TAKE IT ANYMORE! Over 20 years anything that emits pollution — from charcoal barbecues to dry-cleaning chemicals, to gas lawnmowers, to house paints — will be controlled. In their all-out attempt to change human behavior for the sake of the environment, and to reduce pollution over L.A. by 70 percent in the next nine years, the plan's formulators are taking on the icon of Californian culture — the automobile. Day in and day out, in the Los Angeles area, commuters and truckers get into 7.6 million vehicles and drive a total of 376 million kilometers (234 million mi.) on area roads and freeways. All that driving dumps more than 6000 tons of pollution into the air daily.

However it can, the plan intends to break the city's addiction to the automobile. By 1994, California will require all new cars sold in the state to have sophisticated computers to detect pollution-causing problems such as misfiring spark plugs and leaky air conditioners. This move alone is estimated to cut vehicle emissions by 10 percent, and for a cost of only $45 a vehicle.

By 1998, 40 percent of all private automobiles should be running on methanol, and all the district's buses on alternative fuels. By 2007, the plan sees Angelenos scooting around the city in electric vehicles. By 2008, all vehicles sold will have to either run on alternative fuels or be electric.

The plan's originators are convinced that auto makers will rise to the challenge. The California market speaks loudly to the auto industry — one in 20 Americans lives there. At one time, car manufacturers claimed they couldn't lower emissions, but when California began passing air-quality regulations, the percentage of the market it represents literally forced car makers to invent the catalytic converter.

But more than alternative vehicles with alternative fuels, the end goal of the L.A. plan is to wean Angelenos from their sacred cars. There will be stiff fines for companies that don't introduce compulsory car-pooling among employees, a limit on the number of cars each family can own, an end to free parking, government-sponsored work stations set up in suburbs so people won't have to drive into the city, video classes for students in university so they can study at home.

All these policies point in just one direction. Our days as limitless consumers of energy and creators of air pollution in our own private, mobile pollution-machines are coming to an end. The automobile is proving itself to be incompatible with human survival and the well-being of the planet. It destroys our quality of life, the air we breathe, our crops, and our trees with toxic emissions. It destroys the ozone layer. It is responsible for the paving over of our cropland and wilderness. Every time we climb into a car and put our foot on the gas, we're jeopardizing our family's future.

Breaking the habit after a century of use won't be easy. The car has been our identity, our badge of adulthood, our statement to the world of who we are. Will we continue to deny its grave effects? A recent poll showed that 65 percent of Americans would oppose regulations requiring mass transportation and limiting the private use of cars. But they are fighting a losing battle.

Buses are three times more efficient at carrying passengers than is the car, and electric trains the most efficient. A three-car electric train will carry 50 passengers — that's 15 times cheaper than the car in monetary and environmental costs. But the bicycle has them all beat, hands down. For every trip taken by bicycle, there's less fuel used, no pollution created, and more space on the road. More than half of all commuting trips in the United States and nearly three-quarters of all trips in the United Kingdom are 8 kilometers (5 mi.) or less, easy cycling distance for even the most diehard convenience junkie.

What garbage and cars tell us is that the changes we need to make to get us safely into the next century will not be ones of compromise. The planet no longer exists for our exclusive use. It can no longer afford to be at our disposal. Earlier this year a report issued by Environics, the public survey group, on the trends of the 1990s summed us up: "Our surveys indicate that Canadians will continue to structure much of their lives around the pursuit of personal happiness and the enjoyment of an unashamedly materialistic lifestyle. The actual expression of these hedonistic tendencies is at present being tempered by the current economic slowdown, but the values of the 'me' generation will continue to have an overwhelming influence on the decisions Canadians make throughout the 1990s."

Keeping the values of the "me" generation front and center, we're offering compromise solutions to our problems because we're unwilling to be inconvenienced. We will go halfway, but we'll go in the car. We'll drive the empty Perrier bottles to the recycling depot, or throw them in our Blue Boxes, if it's convenient; we'll buy green products with pride — but we'll consume at the same old rates and drive to the store to do it. This new sacred truth of raised consciousness — that we can fit saving the planet into our lifestyle — is as patently false as are the inherited ones we cling to. And it is dangerous. We can't negotiate with a faltering life-support system. We can't buy a future for our children. The only thing we might be able to buy them — if we act now — is a little time: the time it takes to embrace a new way of living, a new worldview.

PART III
TOWARD THE
YEAR 2040

NINE

B U Y I N G
T I M E

In 1986 and 1987 the Environment
and Public Works Committee of the U.S. Senate asked
a dozen leading scientists from the United States and
abroad to testify at extensive hearings. Wallace S.
Broecker, a geochemist at Columbia University, spoke
for many:

> The inhabitants of planet Earth are quietly conduct-
> ing a gigantic environmental experiment. So vast and
> so sweeping will be the impacts of this experiment
> that were it brought before any responsible council
> for approval, it would be firmly rejected as having
> potentially dangerous consequences. Yet the experi-
> ment goes on with no significant interference from
> any jurisdiction or nation. The experiment in ques-
> tion is the release of carbon dioxide and other so-
> called greenhouse gases to the atmosphere.

In recent months, the Beijer Institute in Stockholm, the
U.N. Environment Program, the World Meteorological
Organization, the Worldwatch Institute, the World
Resources Institute, the U.S. Environmental Protection
Agency, the Sierra Club, the Woods Hole Research
Center, and the United Nations Intergovernmental

Panel on Climate Change have outlined strategies for combating global warming. Each group independently concluded that the threat of global warming is so grave that action should be taken immediately.

*Preventing these gases from turning the Earth into a hothouse over the next 50 years means just one thing: we're not going to get away with anything less than "an all-out attack, an all-out assault on the whole process of fossil-fuel combustion everywhere, in order to save the planet,"/says Stephen Lewis. Lewis, former Canadian ambassador to the United Nations, chaired the 1988 international conference on global warming in Toronto. What emerged from the June meeting attended by more than 300 scientists from around the world was the first global scientific consensus that not only are we entering an era of unprecedented climate change, but we're going to have to act in the 1990s to slash carbon dioxide emissions. "It is imperative to act now!" stated the meeting's final communiqué, which called for a worldwide 20 percent decrease in carbon dioxide emissions by the year 2005. An even tougher line has since come out of the U.S. Environmental Protection Agency: in order to stabilize atmospheric concentrations of carbon dioxide at current levels, carbon emissions must be cut by 50 to 80 percent. And in May 1990, the U.N. Intergovernmental Panel on Climate Change (IPCC) warned that if carbon dioxide emissions are not cut by 60 percent immediately, the changes in the next 60 years will be so rapid that nature will be unable to adapt and man incapable of controlling them.

With those figures in mind, Lewis says, the writing is on the wall. The major policy of the Western world has to be "to virtually eliminate dependence on fossil fuels overall, as fast as is humanly possible to achieve."

We all contribute to the problem, demanding electricity, much of which comes from coal-burning generators; driving our cars; heating our homes. Canada and the United States are two of the worst culprits. One North American uses as much commercial energy as two West Germans, three Swiss or Japanese, six Yugoslavs, nine Mexicans or Cubans, 16 Chinese, 19 Malayans, 53 Indians, 109 Sri Lankans, 438 Malians, or 1072 Nepalese. What all that means is that consumption habits of the average North American account for the production of five tons of carbon annually, while people elsewhere in the world account on average for one ton a year.

Not only are we going to have to cut carbon emissions by as much as 80 percent, we're going to have to do it in this decade if we want to avoid a hothouse future; yet, as of this writing, not a single national government has implemented a plan to reduce CO_2 emissions. In fact, we are going in exactly the opposite direction.

A frighteningly large gap is looming between projected growth rates in carbon emissions and the level that atmospheric scientists are convinced is necessary to maintain a climate that will allow humanity to survive. A major study released at the 1989 World Energy Conference in Montreal showed the gap is widening into a chasm — by the year 2020 global industrial emissions of carbon dioxide could be 70 percent higher than current levels, assuming only moderate economic growth. Department of Energy projections in the United States, Canada, and Britain confirm the bad news.

U.S. Department of Energy figures show that rather than a cut of carbon dioxide by 2005, there will actually be an increase of 31 percent over 1985 emission levels. A confidential department of Energy memo in Britain projected an increase of carbon dioxide, in the same period, of 37 percent. In Canada, a report prepared for the provincial and federal Energy ministers in 1989 estimates Canadian CO_2 emissions will rise 49.3 percent by 2005, if current trends continue.

It is going to take personal initiative reinforced by political will to halt the warming trend, but from all appearances, our politicians aren't there yet. While several European countries are formulating programs that would restrict the emission of carbon dioxide and other greenhouse gases to current levels, the U.S. government is calling for more research, unable to accept the strikingly simple answer to the question asked by the American Energy Secretary James Watkins: "Do we have to destroy the industrial base and our economy for world survival?" And while delegates from European countries attending a May 1990 meeting in Bergen, Norway, pushed for an international commitment to limit carbon dioxide emissions, a confidential briefing paper for Canadian delegates to the conference showed that Canada had no intention of seeking progress on global warming: "Canada will not support expected proposals from the Nordic countries for targets and timetables on emission reductions or funding arrangements for developing countries," the Canadian briefing document stated.

Says Michael Oppenheimer, senior scientist with the Environmental Defense Fund in New York, "What they're doing is the height of irresponsibility. They are gambling with our world and they are gambling with our children's world. We know for sure that the greenhouse gases are increasing, and we are certain that this will lead to a significant warming of the Earth."

In Canada, the federal government stalled for more than a year on its green plan for the environment, while the federal Energy minister scuttled a plan for federal-provincial agreement on reducing CO_2 emission levels by 20 percent. This despite a 1989 report commissioned by the federal and provincial Energy ministers that stated such a cutback would result in $100 billion in energy savings for Canadians. And the provincial government of Nova Scotia, claiming that there is no conclusive evidence that carbon dioxide contributes to global warming, gives the go-ahead for construction of a coal-burning power plant that will produce considerably more carbon dioxide than existing plants.

These are the people we elect to office. Normally the decisions they make have short-term consequences. But the stakes are higher in the gamble they're taking with global warming. They are playing with the future of life on this planet. With the odds stacking up against us, indecision itself becomes a deciding factor, and politicians' denial of the warming problem a bluffer's fatal tactic.

"Our leaders have to take this problem seriously," says Doug Scott, conservation director of the Sierra Club. "We need leaders with vision." Scott thinks the public has already grasped in a profound way that the assumptions we've made about the stability of the world we live in are no longer valid; "that the chance to pass on our little bit of earth to our children and our children's children, unchanged, is under great doubt, because change is all around" — everywhere, that is, but where it must be.

The essential change that will enable us to survive into the next century is one that we must make in the next 10 years. "The important rethinking of policy and priorities and behavior will have to come during the 1990s," says Worldwatch's Lester Brown. "If it doesn't, then I think it could be too late — too late in the sense that environmental deterioration and social disintegration could begin to feed on each other." If we maintain the status quo of "business as usual" and persist, both on an individual and on a societal level, in using fossil fuel, we are trading away our children's future. What we must do instead is buy them time.

And that means face a radical change in the way we live. "Carbon dioxide is the exhaling breath of our industrial society," says Albert Gore, chairman of the

U.S. Senate Committee on the Environment. Fossil fuel, like oil and coal, is the engine that drives our civilization and fuels the comforts of our lives.

"We cannot have all the conveniences of our lives — the VCRs, the all-terrain vehicles, the two-car garages, homes filled with all kinds of gadgets — and an easier life, if they're powered by the combustion of fossil fuels," says Doug Scott.

Stephen Lewis concurs: "It's a challenge to the whole economic order. It's a challenge to capitalism as we know it: all the things we take for granted and assume are part of the comfort of Western capitalist society are going to be very seriously challenged." He notes that the obstacles to survival that must be overcome must replace entrenched regional and national economic preoccupations as a political bellwether. Overcoming them will require tough decisions that may, in the short term, be unpopular. Multinational corporations must be brought in line, especially oil companies, who will not want to see their profit margins erode. Consumer patterns must change. Because mandatory restraint fares badly on the hustings, Stephen Lewis thinks Western governments over the next couple of years will do the absolute minimum, even though "this is a desperate race for time."

The crisis seems so immense, and the kind of political inaction that Lewis describes so stultifying, that it's easy to feel powerless to do anything about it. But on this issue we cannot afford to be disenfranchised by our own complacency. You and I are not merely onlookers in the ecological drama that is unfolding before our eyes. We must make our governments understand that cutting carbon dioxide emissions is the number-one priority on our survival list. Energy experts who study the problem are quick to point out that the key is all of us as individuals. "If you break the greenhouse effect down into its simple component parts, it's about energy consumption, it's about agricultural systems, it's about industrialization," says Stuart Boyle of the Association for the Conservation of Energy in England. "The greenhouse effect is us, and therefore the solution is us."

The mandate for survival is broad, far-reaching, and uncompromising. It embraces all aspects of our lives. Eliminating fossil fuels means nothing less than a massive shift from the use of cars to the use of public transit; it means a universal moratorium on use of oil and coal and the substitution of natural gas as a transitional fuel; it means a worldwide adoption of alternative forms of energy, from solar to thermal, and so on. It means large-scale reforestation in developed and developing countries. It means taking dead aim at the cause of global warming, and moving fast, in the next 10 years, to eliminate it. Such strategies, according to the U.S. Environmental Protection Agency, would slow global warming to a rate of between 0.6°C (1.08°F) and 1.4°C (2.52°F) a century — at least 60 percent slower than if we don't do anything.

We have it in our power right now to take the first major step, says George Woodwell, director of the Woods Hole Research Center. "We have the possibility of reducing the use of fossil fuels by something in the order of 50 percent immediately through simple steps in the conservation of energy."

No other approach offers such a large immediate potential for limiting carbon emissions — according to Worldwatch energy estimates, improvements in energy efficiency between 1989 and 2010 could make a three-billion-ton difference in the amount of carbon released to the atmosphere each year.

It can be done. Osage, Iowa, a small community of 3600 people, has got the jump on us all. Surrounded by farmland, Osage doesn't look any different from many small towns in the United States or Canada — but it is. Under the direction of Wes Birdsall, the general manager of Osage Municipal Gas and Electric Company, the town saved an estimated $1.2 million in energy costs in 1988 alone.

Since 1974, when the energy crisis hit, the community has cut its natural gas consumption by 45 percent and Osage's residential heating customers are using 37 percent less gas now during the winter heating months than they were 12 years ago.

In a winter much like a Canadian one, when wind-chill factors can plummet temperatures to -80°C (-60°F) or even to -103°C (-80°F), teacher Ken Swenson's bill for his natural gas furnace is about $50 a year. It's that high, he says, because they run their clothes dryer on natural gas as well. Swenson's home is not small: it has 895 square meters (9634 sq. ft.) of living space.

Wes Birdsall began to preach conservation door to door in 1974. He started with the basics, telling home-owners about insulating walls and ceilings and plugging leaky windows. The Osage Municipal Gas and Electric Company handed out free water-heater blankets to any resident who wanted one. Gradually, Osage residents saw the effect these seemingly small home improvements could have on their utility bills. It became a topic of conversation, it became a hobby, and it became a challenge.

As the townspeople grew more energy aware, Wes Birdsall kept after them, with more tips and with more offers to point out trouble spots. When he offered to give every building in town a thermogram, a test that determines where heat is being lost, more than half of the town's property owners accepted the offer.

Besides the gentle persuasion, there were also more forceful tactics. The utility announced that no new houses could be hooked up to the natural gas line unless they met certain minimum efficiency standards.

Is Osage any different from other towns in the United States or Canada? Besides the amount of gossip about who's doing what to their house and how efficient it'll be, Richard Woodruff, an Osage resident, feels there is something different about Osage. But it's something the town's commitment to energy efficiency has helped to foster: a tremendous sense of community pride and self-reliance. "So many times we look at a problem and think it seems so insoluble, but when each individual takes responsibility, you see there is something you can do."

As the town of Osage, Iowa, goes, so too could the world. Take electricity, for example. Analysts at the American Council for an Energy-Efficient Economy estimate that 35 to 40 percent of U.S. electricity could be saved. In 1988, it took 500 billion kilowatt-hours — the output of 100 standard power plants, or 20 percent of all electricity generated in the United States — to provide lighting to Americans.

The electricity used for all that lighting alone could be cut by 50 percent by replacing normal light bulbs with modern compact fluorescent ones. These bulbs are more expensive, but produce only one-sixth of the carbon dioxide (in terms of energy consumption) of standard light bulbs. An 18-watt screw-in fluorescent bulb produces as much light as a 75-watt incandescent bulb and has a 7500-hour lifetime, 10 times that of an incandescent bulb. Each compact bulb will reduce carbon dioxide emissions from a typical coal-fired power plant by *one ton* over the bulb's lifetime.

Worldwatch figures indicate that commercial buildings could reduce their electricity use for lighting by as much as 75 percent by using similar bulbs and current regulators. The savings add up in concrete terms to the elimination of 40 power plants.

Some of the biggest greenhouse culprits are the American power utilities. According to data compiled by the Electric Power Research Institute in Palo Alto, California, U.S. utilities alone contribute 7.5 percent of the world's carbon dioxide output, mostly from coal-burning power plants. But power utilities are discovering that investing in energy-saving strategies can actually save them the cost of new power plants.

In the United States, some utilities are giving away the high-efficiency light bulbs. Others are operating a rebate program for people who switch to high-efficiency appliances. Some utilities in California not only give a $50 rebate to those who make the switch, but they also haul the old appliance away. Says energy policy analyst Irving Mintzer, "It's worth more than $100 to the utility to not invest in the additional electric-generating capacity that would be required to provide the electricity to operate an inefficient refrigerator in your house."

Figures compiled by the New Democratic Party in Ontario reveal that if Ontario Hydro gave away new energy-efficient appliances to every home in the province, the energy saving would make it unnecessary to build another nuclear plant. It would cost Ontario Hydro up to $7 billion for the appliances; but a new nuclear reactor could cost $17 billion. The logic is inescapable, but it eluded Ontario Hydro, which is in the process of trying to get approval for construction of three new nuclear plants.

Light bulbs and appliances are only the start. Up to 75 percent of the energy used in homes today can be saved, according to statistics compiled at the Rocky Mountain Institute in Colorado. "The tube of caulk is probably the most important efficiency tool that you can work with in the residential sector of housing," says Ted Flannigan, energy director for the institute. "If you add up all the little cracks around the windows and baseboards and looked at all the areas we have thermal loss, it would add up to .09 to .18 square meters (1 to 2 sq. ft.) of leakage in every house in North America."

Advanced building materials can sharply reduce heat loss through windows, doors, and walls. Some superinsulated homes in Minnesota require 68 percent less heat. In Sweden, a world leader in energy efficiency, the saving in some residences is 89 percent. One model home built near New York could even be heated for about 1400 kilowatt-hours. That's about the yearly energy consumption of a refrigerator. Windows have been developed that are equivalent to seven panes of glass. Arthur Rosenfeld, director for Building Science at the Lawrence Berkeley Laboratory in California, is optimistic about efficiency: "If you take superwindows, as we call them, plus lots of insulation, plus mechanical ventilation with heat exchangers, you can build houses all over North America that require very little heat, maybe none, because there's enough heat radiating from the appliances and the people to heat the house . . . so the problem's more or less solved."

However, as powerful a strategy as it is, energy efficiency by itself will not solve the global warming problem. George Woodwell says his calculations demonstrate that carbon dioxide output would have to drop by 75 or 80 percent by the end of this decade just to put global warming on hold; and that means, says Woodwell, "to all practical purposes, that the era of fossil fuels has passed, and it's time to move on to the new era of renewable sources of energy."

The problem is urgent, says Woodwell: the more the Earth warms, the more it will warm. The carbon dioxide that has been dumped into the atmosphere since the Industrial Revolution has already guaranteed that the Earth will warm a degree or two in the next several decades. The crux of the problem is that once started the warming is continuous. "The Earth is not simply moving toward a new equilibrium in temperature, through a brief period of adjustment," says Woodwell. It is entering a period of continuous, progressive, open-ended warming. There is nothing that will hold the increase of greenhouse gases to a doubling, unless we stop their production.

The only way to prevent those gases from even doubling is to move to alternative energy sources that are not carbon based. To date, there are only two available: nuclear energy and renewable energies, like solar power. Natural gas, the transition fuel for the 1990s, contains 44 percent less carbon than oil does, and 75 percent less than coal. But because it still contributes to greenhouse warming, natural gas is seen only as a short-term solution until other energy sources come on stream.

Nuclear energy is held up as a panacea by some governments, including those in Britain and Canada. Both are promising to make it a centerpiece of their future energy policies, and have already invested billions of research and development dollars in nuclear technology. But accidents at Three Mile Island and Chernobyl have made the public wary of nuclear energy. And, of course, there is the problem of nuclear waste.

Environmentalists caution that a move to nuclear energy could mean exchanging one environmental catastrophe for another. Some radioactive waste lasts so long that it might as well last forever, and such options as burying it or encasing it in glass have not been satisfactorily worked out. Plutonium-239 takes 24 000 years to lose half of its radioactivity. Another byproduct of the nuclear process, iodine-129, has a radioactive half-life of 17 million years.

Even discounting the accidents and the worry over what to do with the radioactive waste generated by nuclear plants, energy figures suggest that nuclear is not the route to go. It will not release us from our greenhouse dilemma quickly or cheaply enough.

A study recently completed by the Rocky Mountain Institute concludes that every dollar spent building nuclear-power plants could be seven times as efficacious in diminishing the greenhouse effect if it were invested in improving energy efficiency. Replacing all the world's fossil-fuel-fired power plants with nuclear plants over the next 35 years would cost $144 billion annually and only cause carbon emissions to level off, not decrease, says a 1989 Worldwatch report.

Figures computed by energy analyst Charles Komanoff show that for nuclear power alone to reduce fossil-fuel use by 50 percent, 16 nuclear power plants a week would have to be completed, somewhere in the world, between 1995 and 2020.

The Rocky Mountain Institute had arrived at similar conclusions in the December 1988 issue of *Energy Policy*: "Even if large nuclear plants could be built every one to three days from now until 2025, carbon dioxide emissions would still continue to grow."

Those construction rates seem totally unachievable when you realize that today there are just 94 nuclear plants under construction worldwide — the smallest number in 15 years. The United States put all construction plans on hold in 1979 after the Three Mile Island nuclear accident. According to the Worldwatch *State of the World 1990* report, even if all the plants currently under construction were completed, they would displace only an additional 110 million tons of carbon annually — a tiny fraction of the reduction needed by the end of the decade.

Yet politicians seem to be immune to statistics when it comes to nuclear energy. To them, nuclear plants are monuments to progressive thinking and stand as visible proof of political action. Other energy sources for the post-fossil-fuel era are not as glamorous as nuclear, but they are certainly proving to be more functional:

— Biomass sources of energy include wood, agricultural wastes, and garbage. Already this form of energy supplies about 12 percent of world energy, a figure that reaches as high as 50 percent in some developing countries. While burning of wood and wastes can exacerbate global warming by the release of carbon dioxide, biomass, if properly developed, could be useful.

— Wind energy has been harnessed by Holland for centuries. There, windmills provide power to draw water, grind grain, cut wood. Now scientists have revamped the sturdy, picturesque windmill, turning it into a sleek wind machine designed to capture energy as efficiently as possible right from the air. Today, more than 20 000 electricity-producing wind machines are in use worldwide. It has been estimated that by the year 2030 wind power could provide more than 10 percent of the world's electricity.

— Power plants that burn methane produced in landfill sites are particularly effective at slowing global warming, since they consume a gas with 25 times the greenhouse strength of carbon dioxide. A California study found that one kilowatt-hour of electricity produced this way removes methane equivalent to the carbon released by 10 kilowatt-hours generated by a coal-fired plant.

— Energy derived from water now accounts for 19 percent of the world's electricity. Two-thirds of the unexploited hydropower potential lies in developing countries, but environmental and social problems limit the appeal of large-scale dam projects. Every developing country has its horror story of reservoirs that have flooded valuable farmland, destroyed species, displaced people from their homes, and become a breeding ground for water-borne diseases. But in many cases, small-scale hydro stations can satisfy local needs.

— Geothermal power, which taps underground pressurized hot water for energy, is another alternative.

All these renewable sources will be useful in the new energy world. But to stabilize the climate, carbon emissions have to be cut to two billion tons, about one-third of what we pour into the atmosphere right now, and to that end, the sun is fast becoming the center of our energy universe.

It's hard to imagine powering twentieth-century industry with sunshine, but that is, in fact, the strongest contender for primacy in the new-energy world. Twenty years ago, solar-power enthusiasts were viewed as pie-in-the-sky dreamers. But there is nothing theoretical about solar energy anymore. We are, for all intents and purposes, entering the solar age.

In the California desert northeast of Los Angeles, miles and miles of mirrors are proving just how powerful an energy source the sun can be. Using a system of 1.8 meter (six ft.) high curved mirrors that focus the sunlight onto oil-filled tubes, the Luz company feeds enough energy into Southern California Edison's distribution network to supply the electricity for 270 000 customers. Solar thermal systems like this one convert up to 22 percent of the sunlight they capture into electricity.

The system works by heating the liquid in the tubes to 390°C (734°F). The tubes are immersed in water, which is boiled into steam; that steam is fed through a turbine that turns a generator. The fuel is pure sunlight; the end result is electricity without greenhouse-gas emissions. Natural gas is the backup system for cloudy days.

Though energy experts see a role for these solar mirrors in replacing fossil fuels, most of the future energy bets are riding on solar cells, *solar photovoltaics*, that convert the sun's radiation directly into electricity. They're like the solar cells that run pocket calculators. The Worldwatch Institute calls it "the renewable energy technology likely to advance most rapidly in the years ahead." These solar cells can be placed anywhere, as small units on rooftops or in massive groupings that become desert power plants.

Photovoltaics are finding a receptive audience in the Third World. Six thousand villages in India without access to power grids are getting their electricity from photovoltaics. They run everything from refrigerators to irrigation pumps to lighting. The sun-powered cells are making inroads in North America. The U.S. Coast Guard currently has more than 10 000 photovoltaic systems to run its buoys and remote-communications network. The Canadian National Railway uses the cells to power a communications station in northern Quebec. Right now, worldwide, these solar cells supply electricity to just 15 000 homes, but the U.S. Solar Energy Research Institute estimates that with technological improvements that will come onstream by the late 1990s, photovoltaics are capable of supplying more than half of American electricity by the year 2040.

The Worldwatch Institute paints a compelling picture of a solar-run planet and says the system could be in place just 40 years from now. Solar panels on our roofs will heat all our water for bathing and cleaning needs. Photovoltaic shingles, which are available today, will make homeowners everywhere producers as well as consumers of electricity. These shingles allow the roofing material itself to become a power source. Photovoltaic cells will have evolved to the point where they become energy plants for remote villages all over the world, doing away with the need for long-distance power lines. Mega-power projects will be obsolete, a relic of the past.

The sun's potential for supplying us with cheap and benign energy in the next few decades is unlimited. In fact, its only limitation is the human inability to imagine life without the fossil fuels that now threaten our survival. Both the American and Canadian governments have demonstrated their lack of vision in their funding record for solar research. To replace fossil fuels with solar energy in the coming century, policies have to be shaped now that direct research and development to that end. Federal research funds in the United States for solar energy fell from $348 million in 1981 to only $55 million in 1989. The Canadian government dropped their support from $272 million in 1985–86 to $74 million in 1988–89. These cutbacks are indicative of a halfhearted commitment to the resolution of our fossil-fuel dilemma. The alternatives exist, but the will to implement them is weak. The major barrier is attitude. We have for generations equated bigger with better and most recently with superior, and have empowered our governments to act accordingly. It's time to change the equation.

The solution lies in renewable energies, nothing else. There are no substitutes for the new energy era, no short cuts to the solution for global warming, no quick fixes; in fact even our best attempts to buy time have some caveats attached. The tree-planting story tells us that.

Trees, it turns out, are the greatest carbon-dioxide-filtering system we have on the planet. As they respire, all plants take in carbon dioxide, release oxygen, and store the carbon as wood or fiber. Some scientists are suggesting that trees could be our temporary salvation and could buy us time in the race against global warming. Planting trees could give us decades while we develop and put in place an energy system that doesn't depend on fossil fuels.

Massive new forests covering the globe and acting as gigantic storage vaults for carbon — it's an image that governments and policy makers have exploited to demonstrate that they are dealing with global warming. "Plant a Tree. Cool the Globe" is the slogan of Global Releaf, a reforestation program launched in 1988 by the American Forestry Association. They've set a target of planting 100 million trees in cities and suburbs around the United States by 1992. In his 1990 State of the Union address, U.S. President George Bush announced a program that would increase the nation's forest cover by a billion trees annually for the next 10 years. As grand as the Bush plan sounds, an additional 10 billion trees over 10 years would absorb only 1 to 3 percent of the carbon dioxide produced by human activity in the United States during that same period. Considering that the country produces 1.3 billion tons of carbon as carbon dioxide a year — over 20 percent of the total that enters the global atmosphere — 10 billion trees seem to be a Band-Aid. Last year the prime minister of Australia announced a program to plant one billion trees by the end of the decade. In Britain, the Countryside Commission announced the creation of a new national forest of 30 million trees. And Brazilian scientists released a plan in May 1990 called "Forests for the Environment," which promises to plant 10 billion trees.

Plant some trees and deal with global warming — it has a nice ring to it; it's clean, green, easy to do, a motherhood solution, but unfortunately the problem is not that simple.

Greg Marland of the Oak Ridge National Laboratory in Oak Ridge, Tennessee, was one of the first to look at this possibility, and he thinks the dimensions of the "fix" are staggering: to take five gigatons a year of carbon dioxide from the atmosphere, an amount roughly equal to what we are adding annually, would require between 517 996 860 hectares (1.28 billion acres) and 667 730 250 hectares (1.65 billion acres) of new forests. To put that into perspective: it would be like planting trees over 73 percent of the United States. Marland concludes his analysis with the comment that looking to forests to solve the CO_2 problem is "unrealistic," but he adds an important qualifier: reforestation could play a significant role "as one component among a variety of measures" implemented to address the CO_2 crisis.

The Worldwatch Institute says before we can even address the possibilities of reforestation we have to talk about stopping deforestation. In 1988, 25 percent of the carbon released into the air (one to two billion tons of the total 7.66 billion ton emission) was generated by the felling and burning of forests, mainly in tropical areas. Every ton of carbon released through burning results in 3.7 tons of carbon dioxide. Brazil, for example, is the fourth-largest carbon dioxide emitter in the world, mainly because of the carbon released in the deforestation of its tropical rain forests.

Worldwatch says a combined global strategy of both halving tropical deforestation and of planting the equivalent of 130 million hectares (321 million acres) of trees in developing countries and 40 million hectares (99 million acres) in industrial countries would chop the carbon emissions from human activities by 25 percent of current levels. And those two together, says Worldwatch, are the insurance we need to "slow the pace of warming by several decades, buying precious time to adapt and respond to climate changes in other ways."

The major question is, where would all the new trees go? George Woodwell, who has studied the problem for the Woods Hole Research Center says:

I don't like to be pessimistic, but to store a billion tons of carbon in a forest requires about one million square kilometers (386 130 sq. mi.), perhaps two million square kilometers (772 260 sq. mi.) of new forest. It's hard to find areas on that scale available for reforestation. The trees have to be put into forest that will remain forest and not be damaged by disease, harvested, burned in forest fires, or destroyed by climatic change. Any one of those steps would release carbon back into the air. Planting trees doesn't mean that we naturally have a forest. It took 400 years to grow those monstrous Sitka spruces on the west coast the last time around; it'll take another 400 years to reestablish trees and forests that contain a pool of carbon that's equivalent.

Also, we can't just take wasteland and turn it into carbon-storing forest. Trees have soil requirements for growth, and most of the suitable land in the world is already used for agriculture. Woodwell cites India as an example. One-third of that country is not used in agriculture, but that land is either salinized or rocky and barren, and not capable of supporting plant life.

"I doubt that there are millions of kilometers available, which is what's required globally to reestablish forest," says Woodwell. "It is possible to store carbon by reforestation, but it would be much wiser to stop deforestation now and reduce the use of fossil fuels on a global basis. Planting trees is a stopgap, not a panacea."

The caveats that Woodwell and others place on tree planting as a solution to our global-warming problem just accentuate the fallacy of believing that there is a quick fix for our carbon dioxide problems. As Woodwell points out, nature has too many ways of proving us wrong. What we must do to prevent uncontrollable global warming is evident. Whether we will take the steps necessary is something else again.

It is 1990 and the frog is still sitting in the water and the water is still heating up. The warnings continue, the facts accumulate, but nothing has changed, and everything is still at stake. If it's not warm enough for us yet, and we truly can't perceive the trends, and our politicians are blinded by their own short-term agendas, what will it take to move us to action? Some say a catastrophe, but Stephen Lewis believes that if we allow things to progress that far, it will be too late. "What's at stake at minimum is a loss of human life in unthinkable numbers as drought and famine, a result of the warming process, begin to take their toll." That's at minimum. At maximum, Lewis believes what's at stake is the "civilized planet." We face a breakdown of political stability as desperation grows in Third World countries and resentment and bitterness increase as the West continues to treasure and support its wealth at the expense of the rest of the world. "If we don't take measures, then I suppose it could be the end of the planet as we know it," warns Lewis. "Assuming that we may have underestimated what is already in place and assuming we don't get around to reversing the trend for another 10 or 15 years, that we play at the edges with 20 percent solutions for a number of years while the politicians protect their skins, it may be irreversible."

Arthur Rosenfeld sees that possibility all too clearly:

> We will flirt with carbon dioxide until we have several hot summers in a row, but by then we will have come closer to danger. I think that if we can keep global warming down to something like 4°C (7.2°F), there's a 90 percent chance that all it will cost us is money and discomfort. But there's a 10 percent chance that we will stub our toe on some ecological uncertainty — forests dying too fast, tundra rotting too fast, ocean currents changing, air currents changing — that will set the world off on an instability, and then the warming will be very sudden. We could go cold turkey on fossil fuels and it wouldn't do any good whatsoever, because the carbon dioxide would then be coming from the dying forests. And then we would be talking about whether 10 percent of mankind can survive, not how much it is going to cost us.

If planting trees is a stopgap, and energy efficiency will take us halfway there, and a renewable-energy future still lies just out of reach, then what else can we do to make sure the nightmare world of 2040 doesn't happen? In this matter — the matter of survival — we all have a vote: old, young, rich, middle class, poor, and the truly dispossessed; First World, Second and Third. And we will be casting that vote on behalf of our children, our children's children, and their children after them. We must tell our legislators at every level that they must lead us into this new world, with hope and with vision, and we must speak with one voice:

— If our leaders must legislate us into a new era of renewable energy, then do it.
— If they must redistribute the pie to help Third World countries so we all survive, then do it.
— If we must be legislated out of our cars, then do it.
— If industry must be legislated in order to become environmentally responsible, then do it.

If the United States, which is responsible for more than 20 percent of the carbon dioxide from fossil fuels that threatens us with a hothouse future, lobbies internationally to prevent global reduction quotas, then Americans should tell their government that this is not acceptable. If Canada, which is responsible for more greenhouse-gas emissions per capita than almost any other country in the world, refuses to set even a 20 percent reduction target, then Canadians should tell their government that this is not acceptable; that the Canadian government's slashing of more than 70 percent of the funds for solar research is not acceptable; and if our future means a pact with the Third World for mutually assured survival, and a sharing of the global wealth, then the Canadian government's chopping of the foreign-aid budget by 12.2 percent, with another $1.8 billion to come over the next five years, is not acceptable. Right now the only actions that should make sense to anyone, anywhere on this planet, are those that all our governments take together to get us safely into the next century.

At some time in our distant past the human mind invented an abstract concept — *the future*. Of course the future doesn't exist, except as an idea, as the hypothetical extension of the present. However, by inventing a future, we created *choices* or *options* from which to select the best strategy to ensure our future well-being. That tactic has enabled us to survive to the present.

Now we're being told that the strategies we have chosen for all of history are putting our future at risk. "We are in an era that is so unprecedented that the very fabric of what we've called reason for 2000 or 4000 years is no longer valid or valuable," says Robert Ornstein, co-author of *New World, New Mind*. It is often said that those who don't study the past are condemned to repeat it. But Ornstein thinks it's different now, that the past is no longer "a prologue to the human future." Through thousands of years we've needed disasters to make us change, says Ornstein. "Only measurable catastrophic occurrences make any difference to us. This is what's happened over and over again. . . . But our problem now is that we can't extrapolate from current catastrophic occurrences into the future because those occurrences can actually obliterate life on Earth." And so we must quickly adopt new strategies if we are to survive.

People are capable of transforming their lives overnight. When confronted by an immediate threat, individuals have reordered their priorities and societies have altered direction. We saw it in the Second World War. The United States changed very rapidly after the Japanese attack on Pearl Harbor, as did Canada.

Women left their homes to build aircraft, car factories began turning out tanks, and people adopted a spirit of sacrifice and cooperation to meet the challenge. Paul Ehrlich says we have to believe that what we face now is worse than a million Pearl Harbors, all happening at once.

But we can change direction.

The dramatic political transformation of Eastern Europe demonstrates that fundamental change happens — if people speak out for it, take a stand for it, and mobilize to act for it.

TEN

A   SENSE
OF   PLACE

Peel away the present and peer into the past, back through hundreds of thousands of years. There is a manlike creature moving across a grassy African savannah; near him, a female and a young one. A warning shriek of birds rises around them, and for an instant, they freeze. Then silence, and they flee to the safety of nearby trees.

In the earliest years of human evolution our ancestors knew immediately that danger was at hand. They read the signs and silences of the natural world, and they noted the absences, and they heard clearly the warning cries of the beasts that were their sentinels for survival. They heard, and saw, and they responded at once; and so they made it into the next generation.

For more than a year David Wake made systematic notes in his computer in a file he called "the frog log." Wake, a University of California biologist whose speciality is amphibians, began to observe a disquieting pattern. Everywhere he normally went to research, toads and frogs had either disappeared or their numbers had been dramatically reduced. He wrote to his colleagues, asking them if they had noticed anything unusual, and the reports started to flood in: from Australia — about 20 different amphibian species had been affected; from Halifax — several species are on the wane or haven't been sighted recently, including leopard frogs; from Puerto Rico — three species of miniature frogs have totally disappeared; from the Rocky Mountains — the once-bountiful leopard frog is found now in only four of 33 sites, and the boreal toad in just 10 of 59 areas it had frequented. From Norway, Costa Rica, Central and Northern Europe, the United States, and Canada, the story was the same — all over the world, frogs and other amphibians are disappearing from ponds, swamps, and rain forests, many of the areas untouched by man; precipitous declines were documented, ranging from 50 percent to 90 percent, and some species had gone extinct. At an emergency meeting called in February 1990, at the University of California, researchers admitted that the cause is a mystery (theories ranged from acid rain, to increased ultraviolet light because of ozone depletion, to pesticides, to global warming), but the ramifications are terrifying: "the possibility [exists] that the dying animals are an ominous indicator of human-made environmental problems" — a giant hole is growing in the fabric of life.

We have lost the ability to hear the warning cries of nature, and when, after many biologists notice over several years that something is amiss — first in their own local research terrain, and then, after comparing notes, all over the world — we do not know how to interpret the evidence or how to react, or even if we should react.

We have alienated ourselves from the natural world, and except for a minority of people, including some indigenous peoples, no longer communicate or live in harmony with it. Not long ago, miners took canaries into coal mines with them. When the birds fell over dead, the miners knew there was too much carbon monoxide in the air, and that they had better get out fast. Now, the entire planet is like that mine, the species of the Earth like the canaries. Yet we do not read the signs. If the fish or birds or insects or amphibians fail to show up or to hatch in their usual numbers, we tend to conclude that something odd must have happened locally and temporarily. Few of us would ever think of looking for a global trend or pattern.

We think we no longer need nature. We create economic and religious worldviews that put man's enterprise at the center of the universe and layer it with *sacred truths* that we know now are neither sacred nor true. But the point may be that nature does not need us. There are those who mourn our loss of nature, a loss of the natural beauty we see around us, but the real loss may go unmourned — the real loss may be us. Nature will survive; humanity runs the risk of being written out of the picture. There are scientists who see all life on Earth as a living whole. British biologist James Lovelock calls that concept "Gaia," after the Greek goddess of the Earth. That theory states that the sum total of all living organisms behaves as a single system, that the entire Earth is a living, breathing, self-regulating entity.

"In biological terms, humans provide no essential functions for the survival of other large communities of life forms — save, perhaps, for our own domesticated animals, plants, and parasites," says Timothy Weiskel of the Harvard Divinity School. "If we disappear, it is probable that wheat, rice, cattle, camels, and the common cold virus will not survive in their current forms for very long. But the vast majority of the Earth's organisms can do perfectly well, indeed perhaps thrive . . . without us or our biological associates."

Our sojourn — or the myth of it — as managers of the Earth, following a biblical imperative to have dominion over and act as the stewards of the planet, is at an end. We do not stand supreme. We stand outside. Darwin had it wrong: it is not the *fittest* who will survive, it is the *fit-ins*. Father Thomas Berry, author of *The Dream of the Earth*, calls himself a "geologian," a theologian of the Earth. He has a radical suggestion for recovering our relationship with nature: what we should do is "put the Bible on the shelf for 20 years until we learn to read the book of nature." Berry doesn't believe people can be religious in a man-made world: "Religion emerges out of the grandeur and magnificence of the natural world. Insofar as we diminish the spendor of the natural world, we diminish our experience of the Divine. We can deal morally with suicide, homicide, and genocide; but we commit biocide and geocide — that is, the killing of the life systems of the planet, and the killing of the planet itself — and we have absolutely no moral discipline to deal with it."

As a species, we must negotiate a new contract with nature, one that recognizes that if we are to survive, nature must survive. The most important trait the human species evolved is a large brain from which the mind emerged. It has gifted us with intelligence and foresight, abilities that must not be denied or downplayed. The challenge now is to use that mind for more than just the brute subjugation of the world around us, to bring the human impulse into some kind of balance. The challenge for us is to discover a new humility, a recognition that the overwhelming complexity of nature will never be fully comprehended or controlled. We must accept that we are not the center of the universe, but an integrated part of it, dependent on forces far beyond our control. We must accept from nature, with gratitude, the accommodation of our existence and the succor that enriches it. We must not take more than we need, for such wanton disregard for the natural equilibrium will limit our prospects. It is within our capacity to recognize our place in nature, and to consciously define and accept it. Perhaps the greatest insight that awaits us is that we are not isolated individuals delineated by the surface of our skin. We are connected to all other living organisms through our evolutionary history, a history each of us carries in DNA and shares, through the air, water, and soil, with the entire web of life on Earth. The substance of our species comprises the atoms and molecules of every generation of life since the first cell on Earth and will become compost for all future life forms. The destruction of any part of this web of interconnectedness is the destruction of the destroyer.

Like the earliest human beings whose emerging consciousness and self-awareness posed questions about who they were, how they got here, and where they were going, we must inquire and search for those answers. To renegotiate our contract with nature, to alter our worldview, the party of the first part must be changed from *I* to *we*. The *we* is all of humanity. Our future lies in group survival and group success, not in individual achievement. We must understand that we share the planet with other species that we depend on to survive, that the destiny of all species is a shared one, because to extricate ourselves from the web of life is to perish.

If the human species is genuinely exceptional, it will be shown by what we do in the coming decade. No other species has been able to anticipate limits and deliberately avoid a catastrophe by taking appropriate action ahead of time. We have to face the reality of what scientists are telling us. We have a stake in what happens because our children will inherit what we leave them. But until we truly accept the reality that this planet may soon become unlivable, we'll only make gestures.

If our leaders among government, industry, and workers really see that it's a matter of life and death, a matter of survival, then they will have to act. If the very stuff that we need to breathe, drink, and eat hangs in the balance, can we continue to say that economic growth, profit, material goods, or even political power are the bottom line?

To recognize the crisis that lies ahead and embark on a crash program to develop new values and priorities is an exhilarating challenge. After generations of increasing isolation from the natural world, we are encrusted with ideas and beliefs that have to be scraped away so we can return to the important fundamental questions about why we are here, what constitutes worthy goals, how we define progress. Priorities have to be rediscovered — love, belonging, family, community, sharing, a love of nature.

This is the test for humanity. Will we degenerate into territorial creatures struggling for power, land, and survival, or will we emerge with a new collective image of ourselves as a species integrated into the natural world?

In times of crisis, people have pulled together and forgotten their mistrust and petty rivalries. They've sacrificed and worked to change their lives. There has never been a bigger crisis than the one we now face. And we are the last generation that can pull us out of it. We must act because this is the only home we have. It is a matter of survival.

Introduction

p. 2 **"Something is seriously wrong..."** Paul Ehrlich interview on CBC Radio series "It's a Matter of Survival," broadcast July/Aug. 1989.

p. 3 **"Civilization as we know it..."** Lester Brown interview on CBC "Survival."

p. 4 **"Creating in a single generation..."** U.S. Senator Al Gore interview on CBC "Survival."

One: Beyond Your Worst Nightmare

p. 8 **"Right now, in 2040..."** James Hansen interview on CBC "Survival."

p. 9 **"One of the greatest causes..."** Jodi Jacobson interview on CBC "Survival."

p. 10 **We are already on our way...** Mick Kelly interview on CBC "Survival."

p. 11 **Ten percent carbon dioxide rise since 1983** — "China Leads a New Surge in Output of Greenhouse Gas," *New Scientist*, July 1, 1989, p. 38.

p. 12 **Methane from cattle herds...** Fred Pearce, "Methane: The Hidden Greenhouse Gas," *New Scientist*, May 6, 1989 pp. 37-41.

p. 13 **And if the West Antarctic...** Michael Parfit, "Antarctic Meltdown," *Discover*, Sept. 1989, pp. 39-47.

p. 13 **"We're facing a problem..."** Jim McNeil interview on CBC "Survival."

p. 14 **"We're looking at land loss..."** Stephen Leatherman interview with Jay Ingram, CBC Radio series "Quirks & Quarks," broadcast April 21, 1990.

p. 14   **"It is time to stop waffling..."** James Hansen, "The Greenhouse Effect: Impacts on Current Global Temperature and Regional Heat Waves," statement to U.S. Congress, Senate Committee on Energy and Natural Resources, Hearing, June 23, 1988, 100th Congress, 1st. sess., 1988, pt. 2, 42–49.

p. 15   **Goldilocks theory** is discussed by Stephen Schneider in his book *Global Warming Are We Entering the Greenhouse Century?*, San Francisco: Sierra Club Books, 1989, p. 19.

p. 16   **"We already know enough..."** Stephen Schneider interview on CBC "Survival."

p. 17   **"If we could say the world..."** E. O. Wilson interview on CBC "Survival."

p. 17   **"It would be sheer..."** Stephen Schneider interview on CBC "Survival."

p. 18   **"Fifty-one years ago..."** Joel B. Smith future scenario on CBC "Survival."

p. 18   **Potted plant story and tree movement with climate change:** Michael Soule's work as discussed by Andy Dobson, Alison Jolly and Dan Rubenstein, "The Greenhouse Effect and Biological Diversity," *Trends in Ecology and Evolution*, Vol. 4, No. 3, pp. 64–68, March 1989.

p. 20   **"Canada was less damaged..."** John Last future scenario on CBC "Survival."

p. 23   **The Thomas Midgley story** is found in Editorial Staff Report, "Chemical Advances Reported in Atlanta Sessions," *Chemical and Metal Engineering*, Vol. 37, No. 5, 1930, p. 286.

p. 24   **Rowland and Molina story** described by Randall Black in "Propelled into Controversy," *UCI Journal*, Nov./Dec. 1986.

p. 24   **"Our immediate reaction..."** Sherry Rowland quote in story by Randall Black, "Propelled into Controversy," *UCI Journal*, Nov./Dec. 1986.

p. 25   **"Orchestrated by the Ministry..."** quote from article written by Lanie Jones, "He sounded the alarm, paid heavy price," *L.A. Times*, July 14, 1988.

p. 26   **Story of sitting on CFC depletion findings** told by F. Sherwood Rowland, "Can We Close the Ozone Hole?" *Technology Review*, Aug./Sept. 1987, pp. 51–58.

p. 27   **"Even if we had..."** Sherry Rowland, AP-CP London, Ozone Meeting, March 5, 1989.

p. 27   **Ozone recovery delayed...** NASA researchers announced in April of 1990. Michael J. Prather and Robert T. Watson, "Stratospheric Ozone Depletion and Future Levels of Atmospheric Chlorine and Bromine," *Nature*, Vol. 344, April 19, 1990, p. 729.

p. 27 **Most cfcs remain in atmosphere ...** longevity information found in "The Ozone Layer" Atmospheric Environment Services Fact Sheet, Environment Canada, March 1988.

p. 27 **Problems with cfc replacements** discussed in "cfc users faced with finding substitutes," *The Economist*, as reprinted in *The Globe and Mail*, April 16, 1990.

p. 28 **A report released at the London meeting** Philip Shabecoff, "Scientists report more deterioration in Earth's ozone layer," *The New York Times*, June 24, 1990.

p. 28 **One percent decrease in ozone** will lead to a 3 percent increase in skin cancer worldwide... U.N. Environment Program draft report, "Environmental Effects Panel Report," August 1989.

p. 28 **Effects of ozone depletion on health and food supply:** Jan Sinclair, "Ozone Loss Will Hit Health and Food, Says U.N. Study," *New Scientist*, Feb. 3, 1990, p. 27.

p. 29 **"Australia was one..."** Ian Lowe interview conducted by Science Unit of Australian Broadcasting Corporation, Nov. 27, 1989.

p. 31 **Standoff between White House Chief of Staff John Sununu** and EPA Administrator William Reilly discussed by Leslie Roberts, "Global Warming: Blaming the Sun," *Science*, Vol. 246, Nov. 24, 1989, pp. 992–993.

p. 32 **"Essentially wishes away global warming..."** Ibid.

p. 32 **Report that attacked global warming predictions,** by William Nierenberg, Robert Jastrow, and Frederick Seitz, "Scientific Perspectives on the Greenhouse Problem," George C. Marshall Institute, Washington, D.C., 1989.

p. 32 **"That 0.5 to 0.6 degree warming we've had..."** Stephen Schneider as quoted by Leslie Roberts, "Global Warming: Blaming the Sun," *Science*, Vol. 246, Nov. 24, 1989, pp. 992–993.

p. 32 **"Sununu is holding the Marshall..."** Stephen Schneider as quoted in *Science* article.

p. 33 **"A lot of people say..."** Stephen Schneider interview on CBC "Survival."

p. 34 **The "magic thermostat syndrome"** discussed by Ralph Cicerone, personal communication.

p. 35 **"When I look back..."** Jeremy Leggett future scenario on CBC "Survival."

p. 36 **"We've already made a commitment..."** George Woodwell interview on CBC "Survival."

p. 36 **Everything we now take for granted ...** discussed by Doug Scott of the Sierra Club on CBC "Survival."

p. 37 **Books and reports mentioned are:** *Global Warming, Are*

We Entering the Greenhouse Century?, by Stephen H. Schneider, San Francisco, Sierra Club Books, 1989; *Living in the Greenhouse* by Ian Lowe, Australia, *Scribe*, 1989; Joel B. Smith is one of the authors of the EPA's draft report to Congress, Policy Options for Stabilizing Global Climate, Feb. 1989; Jodi Jacobson is a senior researcher on the Worldwatch Institute's *State of the World Reports* published annually by W. W. Norton & Co., New York/London.

Two: How Did We Come to This?

p. 39   **Boiled frog syndrome story** told by Robert Ornstein on CBC "Survival."

p. 44–46   **The Australia story** was told on the CBC Radio science series "Quirks & Quarks," on Jan. 23, 1988, on the occasion of Australia's 200th Anniversary. Eric Rolls tells the story poignantly in his book *They All Ran Wild*, London/Sydney/Melbourne, Angus & Robertson, 1984; Australia today discussed by Jeffrey C. Rubin, "Regreening a Scorched Land," *Time*, Jan. 22, 1990, p. 54.

p. 47–48   **Easter Island story:** John Flenley, botanist at Dept. of Geography, University of Hull, England, interview with Jay Ingram, host of "Quirks & Quarks," Feb. 7, 1987.

p. 48   **The coastal rain forest ...** John C. Ryan, "Timber's Last Stand," *Worldwatch*, Vol. 3, No. 4, July/Aug. 1990, p. 27.

p. 48   **Alaska's Tongass forest** discussed in article by Martin Walker, "Northern Plights," *The Guardian*, Jan. 12, 1990.

p. 49   **Steller's sea cow extinction story** told by Bill Sargeant, professor of Geological Sciences, University of Saskatchewan, Saskatoon, "Quirks & Quarks," April 5, 1986.

p. 50   **In eight brief years ...** "Will a ban on ivory save the elephant?" *The Economist* as reprinted in *The Globe and Mail*, July 8, 1989.

p. 52   **"More money is spent in the ..."** E. O. Wilson, personal communication.

Three: Nature Is Infinite

p. 57   **"Sooner or later every one ..."** Jacob Bronowski quoted from "Biography of an atom — And the universe," *The New York Times*, Oct. 13, 1968.

p. 59   **Mexico City pollution** has been the subject of many newspaper and magazine articles but particularly vivid accounts can be found in these newspaper stories: Linda Hossie, "Pollution killing thousands in Mexico City," *The Globe and Mail*, Nov. 23, 1988; Bernd Debusmann, "Birds die, Mexico City wonders: who's next," *Toronto Star*, March 8, 1987; Dave Todd, "Mexico City's filthy air blamed for retardation," *Toronto Star*, Feb. 5, 1989; Larry Rohter, "Mexico City's filthy air, world's worst, worsens," *The New York Times*, April 12, 1989.

p. 59   **1989 U.N. Environment Program report on Mexico** discussed by Larry Rohter, "Mexico City's filthy air, world's worst, worsens," *The New York Times*, April 12, 1989.

p. 60   **Respiratory and intestinal infections** are the primary killers of Mexico City's children: Dr Anmaid Toledo, a pediatrician at the National Medical Center; Linda Hossie, "Pollution killing thousands in Mexico City," *The Globe and Mail*, Nov. 23, 1988.

p. 60   **Diplomats advised not to take children to Mexico City:** Larry Rohter, "Mexico City's filthy air, world's worst, worsens," *The New York Times*, April 12, 1989.

p. 60   **One of every 100 ...** Geneticist Dr. Antonio Velazquez Arellano quoted by Dave Todd, "Mexico City's filthy air blamed for retardation," *Toronto Star*, Feb. 5, 1989.

p. 60   **Fetal exposure to levels of lead ...** "Lead and child development," J. Michael Davis and David J. Svendsgaard, *Nature*, Vol. 329, Sept. 24, 1987, pp. 297–300. The authors say: "Apparently then, under some conditions the fetus may be exposed to higher levels of lead than are indicated by the mother's blood-lead concentration: indeed, the fetus may even act as a 'sink' for the mother's body burden, as was first suggested by observations of women employed in British white lead factories around the turn of the century. Some of these women claimed that by becoming pregnant they protected their own health through a transfer of lead to the fetus."

p. 60   **Mexico City comes to grips with its pollution ...** Steve Nadis, "Mexican Cleanup," in *Trends, Technology Review*, Nov./Dec. 1989, pp. 10–11.

p. 60   **"Slight improvement in 10 years..."** U.S. Embassy study quoted by Larry Rohter, "Mexico City's filthy air, world's worst, worsens," *The New York Times*, April 12, 1989.

p. 61   **Survey of polluted air around the world** found in following articles: "City dwellers breathe bad air, study says," Associated Press, Geneva, in *The Globe and Mail*, Sept. 16, 1988; Larry Tye, "East bloc's dirty little secret," *Toronto Star*, Jan. 13, 1990.

p. 61   **"Every 24th disability and every 17th death..."** Hilary F. French, "Clearing the Air," *State of the World 1990*, New York/London, W. W. Norton & Co., 1990, pp. 98–118.

p. 61   **American pollution data —** "100 million Americans breathing polluted sir," Associated Press, Washington, *The Globe and Mail*, March 23, 1989; Philip Shabecoff, "Air poisons called threat to public," *The New York Times*, March 23, 1989.

p. 61   **Ninety percent of all Americans ...** Econews ... Body Weight ... *Greenpeace*, Vol 15, No. 4, July/Aug. 1990. p. 7.

p. 61   **"Our record stinks on air pollution..."** "Our record stinks on air pollution, report charges," Michael Hanlon, *Toronto Star*, May 26, 1989.

p. 62   **"The storm is going to come..."** Ralph Nader interview CBC "Survival."

p. 64   **"If my trees are dying..."** Michael Herman unbroadcast portion of interview for CBC "Survival."

p. 64   **"Trees cause pollution"** quote from President Reagan while on the campaign trail on Sept. 10, 1980. In Steubenville, Ohio, on Oct. 10, 1980, Mr. Reagan told his listeners: "I know Teddy Kennedy had fun at the Democratic convention when he said that I said that trees and vegetation caused 80 percent of the air pollution in this country. Well, he was a little wrong about what I said. I didn't say 80 percent. I said 92 percent — 93 percent, pardon me. And I didn't say air pollution, I said oxides of nitrogen. Growing and decaying vegetation in this land are responsible for 93 percent of the oxides of nitrogen."

p. 65   **U.S. Justice Dept. lawyers argue American industrial responsibility** for Canadian acid rain "entirely speculative": Michael Weisskopf, "U.S. in court, doubts theory of acid rain," *The Washington Post*, Nov. 28, 1989.

p. 65   **"That's only going to..."** Arch Jones unbroadcast portion of interview for CBC "Survival."

p. 65   **Lester Brown's story** of the State of the World's first acid-rain report found in "The State of the World: An Interview with Lester Brown," *Technology Review*, July 1988, p. 55.

p. 65   **Ninety percent of Germany's trees** to go from acid rain — Bernhardt Ulrich quoted by Geoffrey Lean, "Forest without trees signals bleak future," *The Observer*, March 9, 1986.

p. 66   **14 000 lakes dead, 150 000 more at risk in Canada...** J. R. M. Kelso et. al., "Acidification of Surface Waters in Eastern Canada and its Relationship to Aquatic Biota," Executive Summary, Dept. of Fisheries and Oceans, 1986, p. 34.

p. 67   **Northern breast-milk study:** Eric Dewaily, "Proposal for a Quality Surveillance Program of Breast Milk in Northern Quebec," Département de santé communautaire du CHUL, Centre de Toxicologie du Québec, Jan. 1989, p. 12.

p. 68   **"The outpourings of the..."** Dennis Patterson, unbroadcast portion of interview for CBC "Survival."

p. 68   **Northern Quebeckers warned** not to eat deer and moose because of cadmium contamination: "Cadmium Contamination, Beware of Deer and Moose Organs," Loisir, Chasse et Pêche, Quebec government pamphlet, 1988.

p. 69 **"The very survival of our people ..."** Dennis Patterson unbroadcast portion of interview for CBC "Survival."

p. 69 **Giving up Inuit diet** would result in annihilation of entire people... Lorene Gemmill, unbroadcast portion of interview for CBC "Survival."

p. 69 **"Long time ago ..."** Pauloosie Kooneeloosie interview on CBC "Survival."

p. 70 **"It is ironic ..."** Mary Kaye May unbroadcast portion of interview for CBC "Survival."

p. 71 **North Sea pollution** — "Where the rivers pollute the sea," *New Scientist*, December 11, 1986, p. 20.

p. 71 **The pollution of the coastal waters of North America** is documented in a number of articles: Tom Morganthau with Mary Hager, Lisa Brown, Ted Kennedy, Lisa Drew, "Don't Go Near the Water," *Newsweek*, Aug. 1, 1988, pp. 42–47; Jon Ferry, "Polluted waters tarnish British Columbia's pristine image," Reuter, May 30, 1989; Kathleen Kenna, "Victoria B.C. sits on the edge of a giant toilet," *Toronto Star*, Dec. 3, 1989.

p. 72 **Algae becoming a monster:** Clarice Yentsch, Research Scientist with the Bigalow Laboratories for Ocean Sciences in West Booth Bay Harbour, Maine. She has been studying red tides for the past 15 years.

p. 73 **Hong Kong had never...** Dr. Ken Smayda, Professor of Oceanography, University of Rhode Island, personal communication.

p. 73 **When a beluga dies...** from story by Andre Picard, "Wildlife bear graphic scars of chemical abuse of waterway," *The Globe and Mail*, Aug. 17, 1989.

p. 73 **Women not yet past childbearing age** should not eat fish from Great Lakes ... Theo Colburn et al., "Great Lakes, Great legacy," Conservation Foundation and Institute for Research on Public Policy, Jan. 1990, p. 180.

p. 73 **History of the exploitation of the bounty of the seas** found in Farley Mowat's book *Sea of Slaughter*, Toronto, McClelland & Stewart, 1984, pp. 192 and 195.

p. 75 **"We can't count the fish..."** Chris Campbell, unbroadcast portion of interview for CBC "Survival."

p. 75 **"A man that's fishin'..."** Harold Barrett interview on CBC "Survival."

p. 76 **Cod catch should be reduced by 50 percent** — Independent Review of the State of the Northern Stock, submitted by Dr. Leslie Harris, May 15, 1989, pp. 33–39.

p. 76 **"We are at the edge..."** Leslie Harris, personal communication.

p. 76 **"It's renewable but..."** Chris Campbell, unbroadcast portion of interview for CBC "Survival."

p. 77 **"The wall of death"** — drift nets are discussed in a number of articles: Stephen Nicol, "Who's counting on krill?" *New Scientist*, Nov. 11, 1989, pp. 38–41; Ian Anderson, "Driftnet peril moves to South Pacific," *New Scientist*, June 17, 1989, p. 35; Philip Shabecoff, "Huge drifting nets raise fear for an ocean's fish," *The New York Times*, Mar. 21, 1989; Greenpeace Action, Ocean Ecology 1989; Mark Nichols, "An Alarming Catch," *Maclean's*, June 12, 1989 p. 49; Clive Gammon, "A sea of calamities," *Sports Illustrated*, May 16, 1988, p. 48; editorial by Jim Coe, *People*, May 15, 1989.

p. 78 **"We have some claim as..."** Leslie Harris, personal communication.

p. 78 **"I guess when you live..."** Brian Walsh, unbroadcast portion of interview for CBC "Survival."

p. 78 **Draining groundwater sources:** Peter Rogers, "Water Not As Cheap As You Think," *Technology Review*, Nov./Dec. 1986, p. 31; "Water in America: Colourless, odourless, tasteless — priceless," *The Economist*, Oct. 4, 1986, pp. 35–38.

p. 79 **Canadian statistics on groundwater use** — John A. Cherry, Chapter 14, "Canadian Aquatic Resources," *Cdn. Bulletin of Fisheries and Aquatic Sciences*, 215, p. 387; "Water 2020: Sustainable Use for Water in the 21st Century," Science Council of Canada Report 40, June 1988, pp. 11–12.

p. 79 **Grim picture of water shortages worldwide:** Sandra Postel, "Saving Water for Agriculture," Worldwatch *State of the World 1990*, New York/London, W. W. Norton & Co., 1990, pp. 39–58.

p. 80 **Rise in temperature will lead to drop in water supplies:** Sandra Postel, "Water for Agriculture: Facing the Limits," Worldwatch Paper 93, Dec. 1989, p. 32.

p. 80 **"I think a worst-case..."** David Brooks, personal communication.

p. 80 **"Dreaming up schemes like..."** "Water in America: Colourless, odourless, tasteless — priceless," *The Economist*, Oct. 4, 1986, p. 37.

p. 81 **Statistics on North American water wastage:** Peter Rogers, "Water: Not as Cheap as You Think," *Technology Review*,

Nov./Dec. 1986, p. 41; "Water: The Potential for Demand Management in Canada," Science Council of Canada Report, 1988, p. 5.

p. 82 **California water shortage** described by Robert Reinhold, "Rain alone won't ease latest drought in West," *The New York Times*, April 16, 1990.

p. 82 **"Societies to adjust human numbers ..."** Sandra Postel, "Saving Water for Agriculture," Worldwatch *State of the World 1990*, New York/London, W. W. Norton & Co., 1990, pp. 57–58.

p. 83 **"And in our triumph ..."** Paul Ehrlich interview CBC "Survival."

p. 83–84 **Worldwide soil statistics** found in: "The Challenge of Hunger in Africa: A Call to Action," The World Bank, Dec. 1988, p. 3; Katharine Forestier, "The Degreening of China," *New Scientist*, July 1, 1989, p. 52; J. Dumanski, D. R. Coote, G. Lucluk, and C. Lok, "Soil Conservation in Canada," *Journal of Soil and Water Conservation*, July/Aug. 1986, Vol. 41, No. 4, pp. 204–209; A Growing Concern: Soil Degradation in Canada, Science Council of Canada, Council Statement, Sept. 1986, p. 9.

p. 84 **"The average New Yorker ..."** Paul Ehrlich unbroadcast portion of interview for CBC "Survival."

p. 84 **"My guess is that the issue ..."** Lester Brown interview on CBC "Survival."

p. 85 **"Two consecutive years ..."** "World Grain Situation and Outlook," United States Dept. of Agriculture, Circular series, FE 10-89, Oct. 1989.

p. 85 **Grain safety net gone:** Lester R. Brown, "Feeding Six Billion," *Worldwatch*, Sept./Oct. 1989, pp. 32–40.

p. 86 **Canadian farmers quit agriculture:** Geoffrey York, "Drought on prairies, low grain prices prompt 24,000 to quit their farms," *The Globe and Mail*, June 27, 1988.

p. 86 **1989 U.S. grain harvest falls short:** U.S. Dept. of Agriculture Report, Oct. 1989.

p. 86 **"If we look 10 years hence ..."** Mick Kelly interview on CBC "Survival."

p. 86 **Predictions for heat-reduced harvests:** Lester R. Brown, "The Grain Drain," *The Futurist*, July/Aug. 1989, pp. 9–16; "Report warns Saskatchewan of dry future," *The Globe and Mail*, June 23, 1988.

p. 87 **Hunger statistics:** The Cyprus Initiative, Hunger in the World, President's Report to the Fifteenth Ministerial Session, Cairo, Egypt, May 22–25, 1989; World Hunger Fifteen Years After the World Food Conference: The Challenges Ahead, pp. 1 & 4, U.N.

World Food Council; CBC "Survival."

p. 87 **Grain prices up:** The Cyprus Initiative against Hunger in the World, President's Report to the Fifteenth Ministerial Session, Cairo, Egypt, May 22–25, 1989, U.N. World Food Council, p. 10; Lester R. Brown, "Feeding Six Billion," *Worldwatch* Sept./Oct. 1989, pp. 32–40.

p. 87 **"There are some people..."** Paul Ehrlich interview on CBC "Survival."

p. 88 **"If the United States were to experience..."** Lester Brown interview on CBC "Survival."

p. 89 **"The most productive applications..."** Norman Borlaug quoted by Lester R. Brown and John E. Young, "Feeding the World in the Nineties," Worldwatch Institute, *State of the World 1990*, New York/London, W. W. Norton & Co., 1990, p. 69.

p. 89 **The Alan Shawn Feinstein World Hunger Program study** quoted by Paul R. Ehrlich, Gretchen C. Daily, Anne H. Ehrlich, Pamela Matson and Peter Vitousek, "Global Change and Carrying Capacity: Implications for Life on Earth," The Stanford Institute for Population and Resource Studies, Working Paper Series #0022, p. 9.

p. 90 **"Right now, if North Americans..."** Lester Brown personal communication.

p. 90 **The time may have come...** Lester Brown, "Feeding 6 Billion," *Worldwatch*, Sept./Oct. 1989, pp. 32–40.

Four: Go Forth and Multiply...

p. 91 **"We're up against the wall..."** E. O. Wilson interview on CBC "Survival."

p. 91 **"The farmers of the world..."** Paul Ehrlich interview on CBC "Survival."

p. 92 **"The battle to feed..."** Paul Ehrlich and Anne Ehrlich, Prologue, *The Population Bomb*, New York, Ballantine Books, 1968.

p. 92 **"The green revolution..."** Bill Rees unbroadcast portion of interview for CBC "Survival."

p. 93 **"If this planet..."** E. O. Wilson unbroadcast portion of interview for CBC "Survival."

p. 94 **It took 10 000 human lifetimes...** Robert Ornstein unbroadcast portion of interview for CBC "Survival."

p. 94 **Those are the concerns...** Melanie Phillips, "Simply dying to be counted," *The Guardian*, Feb. 23, 1990.

p. 95 **"In human evolution..."** Kingsley Davis, Population Special, Bulletin of the Atomic Scientists, April 1986.

p. 97 **In the slums of Calcutta...** Alan B. Durning, "Poverty and the Environment: Reversing the Downward Spiral," World-watch Paper 92, Nov. 1989, p. 31.

p. 97 **The "demographic trap"...** Lester Brown and Jodi Jacobson, "Our Demographically Divided World," Worldwatch Paper 74, 1986.

p. 97 **"Rapid population growth..."** Lester Brown and Jodi Jacobson, "Our Demographically Divided World," Worldwatch Paper 74, 1986, p. 5.

p. 98 **"Forget plateaus..."** Phillips Cutright quoted in *Taking Population Seriously*, Lappé and Schurman, London, Earthscan, 1989, p. 8.

p. 98 **"Essentially the choice..."** Lester Brown, "Slowing Population Growth," *Worldwatch*, Jan./Feb. 1989, p. 17.

p. 98 **"There is no question..."** Paul Ehrlich, personal communication. Ehrlich goes on to say in his book *The Population Explosion*, New York, Simon and Schuster, 1990, p. 17: "We shouldn't delude ourselves: the population explosion will come to an end before very long. The only remaining question is whether it will be halted through the humane method of birth control, or by nature wiping out the surplus. We realize that religious and cultural opposition to birth control exists throughout the world; but we believe that people simply don't understand the choice that such opposition implies."

p. 98 **The National Audubon Society...** Patricia Baldi; Population Expert Audubon Society, personal communication.

p. 99 **"These right-to-lifers..."** Lester Brown, personal communication.

p. 99 **"All the world's..."** Population Crisis Committee, "1990 Report on Progress Towards Population Stabilization," Washington, Feb. 26, 1990.

p. 99 **In April, the American Assembly...** Leonard Silk, "Economic scene: a global program for environment," *The New York Times*, April 27, 1990.

p. 100 **In November 1988...** Paul R. Ehrlich and Anne H. Ehrlich, *The Population Explosion*, New York, Simon and Schuster, 1990, p. 18.

p. 100 **In May 1990...** Clyde Haberman, "Pope in Mexico, Assails Birth Control," *The New York Times*, May 11, 1990.

p. 100 **"We're in a situation..."** Lester Brown, personal communication.

p. 100   **"It's very important to remember..."** Paul Ehrlich
unbroadcast portion of interview for CBC "Survival."

p. 101   **"These modern messages..."** Aboyami Fajobi, executive
director of Planned Parenthood federation of Nigeria, People
News-Features, "Nigeria Sets Target of Four Children Per Woman," International Planned Parenthood, London, May 1988, p. 2.

p. 102   **People in the rural Third World...** John Caldwell,
(*Theory of Fertility Decline*, New York, Academic Press, 1982),
quoted by Betsy Hartmann, *Reproductive Rights and Wrongs*, New
York, Harper and Row, 1987, p. 7.

p. 102–103   **A World Bank study...** "World Development Report
1984" by Oxford University Press, New York, p. 52.

p. 103   **"Third World parents..."** Frances Moore Lappé and
Rachel Schurman, *Taking Population Seriously*, London,
Earthscan, 1989, p. 22.

p. 103–104   **"Rapid population growth..."** Ibid., p. 54.

p. 105   **Kerala is poor...** Richard Franke and Barbara Chasin
also describe the success of Kerala in "Development without
Growth: The Kerala Experiment," *Technology Review*, April 1990,
pp. 42–51.

p. 105   **"With these few facts..."** Lappé and Schurman, p. 59.

p. 106   **"In the 1990s..."** James Grant UNICEF quoted by Richard
Franke and Barbara Chasin in "Development without Growth: The
Kerala Experiment," *Technology Review*, April 1990, p. 51.

p. 106   **"The most common misperception..."** Paul Ehrlich
quoted in "Rich people spoiling Earth, ecologist says," AP, *Toronto
Star*, April 6, 1990.

p. 107   **"Should sub-replacement..."** A. Romaniuc, "Current
Demographic Analysis;" Fertility in Canada: From Baby-Boom to
Baby-Bust, Demography Division, Statistics Canada, Ministry of
Supply and Services, June 1987, p. 111.

p. 108   **Romania's baby-making story** discussed by Jodi
Jacobson, "Abortion in a New Light," *Worldwatch*, March/April
1990, Vol. 3, No. 2, p. 33.

p. 108   **We've also seen another...** David Hatter, "Quebec's baby
incentive following in barren footsteps," *The Financial Post*, May
16, 1988.

p. 109   **"The issue is no longer..."** Lester Brown, personal communication.

Five: ... And Dominate the Earth

p. 111   **"The Earth exists..."** Ari L. Goldman, "Focus of Earth
Day should be on man, cardinal cautions," *The New York Times*,

April 23, 1990.

p. 111 **Scientists are rarely known...** E. O. Wilson interview on CBC "Survival."

p. 112 **Almost .60 hectare of rain forest...** Philip Shabecoff, "Loss of tropical rain forests is found much worse than was thought," *The New York Times*, June 7, 1990.

p. 112 **A mountain ridge in Peru...** E. O. Wilson unbroadcast portion of interview for CBC "Survival."

p. 113 **What happens in them...** Randy Hayes interview on CBC "Survival."

p. 113 **The plant family...** E. O. Wilson unbroadcast portion of interview for CBC TV "The Nature of Things," recorded Jan. 12, 1989.

p. 114 **There's a moth...** Michael Robinson unbroadcast portion of interview for CBC "Survival."

p. 114 **That's something that...** Jose Goldemberg, unbroadcast portion of interview for CBC "Survival."

p. 115 **Computer simulations published...** J. Lean and D. A. Warrilow, "Simulation of the Regional Climatic Impact of Amazon Deforestation," *Nature*, Vol. 342, Nov. 23, 1989, p. 411.

p. 115 **If the forests are destroyed...** Jose Lutzenberger quoted in article by Pat McNenly, "Destruction of rain forests 'catastrophic' for climate," *Toronto Star*, June 22, 1988.

p. 115 **"We should never forget..."** E. O. Wilson interview CBC TV "The Nature of Things," recorded Jan. 12, 1989.

p. 115 **Wilson discovers a new...** E. O. Wilson, unbroadcast portion of interview for CBC "Survival."

p. 116 **It's research like Wilson's ...** Elizabeth Royte quotes E. O. Wilson in "The Ant Man," *The New York Times Magazine*, July 22, 1990, p. 21.

p. 116 **One hectare of Peruvian rain forest...** Edward C. Wolf, "Avoiding a Mass Extinction of Species," Worldwatch Institute, *State of the World 1988*, New York/London, W. W. Norton & Co., 1988, p. 106.

p. 116 **In a single species...** E. O. Wilson interview on CBC "Survival."

p. 117 **Wilson relates a ...** Elizabeth Royte, "The Ant Man," *The New York Times Magazine*, July 22, 1990, p. 39.

p. 117–118 **Description of the destruction of the world's rain forests** can be found in the following articles: James Brooke, "Coca fields replace rain forests as U.S. demand fuels production," *The Globe and Mail*, Aug. 14, 1989; Striking a Balance: The Environmental Challenge of Development, The World Bank, Sept. 1989, p. 12; Debbie Macklin, "World Bank wields its power to preserve in Madagascar," *New Scientist*, Aug. 25, 1988, p. 27.

p. 118 **In 1831, Charles Darwin...** Edward C. Wolf, "Avoiding a Mass Extinction of Species," Worldwatch, *State of the World 1988*, New York/London, W. W. Norton & Co., 1988, p. 104.

p. 119 **There are plants everywhere...** Stephen Price interview on CBC "Survival."

p. 120 **One small man or woman...** Randy Hayes interview on CBC "Survival."

p. 121 **For weeks and weeks there wasn't...** Stephan Schwartzman, unbroadcast portion of interview for CBC "Survival."

p. 121 **"A green paradise that could..."** E. O. Wilson interview on CBC "Survival."

p. 121 **"The destruction of the..."** Steven Price interview on CBC "Survival."

p. 121 **It is estimated that...** "How Brazil subsidizes the destruction of the Amazon," Economics Focus, *The Economist*, March 18, 1989, p. 91.

p. 122 **The site of some...** Eugene Linden, "Playing with Fire," *Time*, Sept. 18, 1989, pp. 62–68.

p. 122 **"We cannot simply..."** Enio Cordeiro interview on CBC "Survival."

p. 122 **The dreams of developing...** Marlise Simons, "Brazil wants its dams, but at what cost?" *The New York Times*, March 12, 1989.

p. 123 **"The most serious threat..."** Hilgard O'Reilly Sternberg, quoted by Kelly Toughill, "Amazon rainforest could be wiped out," *Toronto Star*, Nov. 5, 1989.

p. 124 **Approximately a quarter of...** Peter Raven, personal communication. Raven adds: "We may start with the assumption that more than half of the total biological diversity in the world, or at least three-quarters of the diversity in the tropics, occurs in regions where the vegetation is already destroyed or very likely to be destroyed over the next 25 years. For this reason, it appears likely that approximately a quarter of the biological diversity existing in the mid-1980s, will vanish during this quarter century."

p. 124 **"All predictions are..."** Randy Hayes interview on CBC "Survival."

p. 124 **"People have this idea..."** Hilgard O'Reilly Sternberg, personal communication.

p. 125 **That shocking waste...** Charles M. Peters, Alwyn H. Gentry and Robert O. Mendelsohn, "Valuation of an Amazonian

Rain Forest," *Nature*, Vol. 339, June 29, 1989, pp. 655–656.

p. 125 **Long before the white man...** Stephan Schwartzman, unbroadcast portion of interview for CBC "Survival."

p. 126 **Tribal peoples can tell...** Judith Gradwohl unbroadcast portion interview CBC "Survival." Judith Gradwohl and Russell Greenberg look at how the rain forest can be used in a sustainable way in their book *Saving the Tropical Forests*, London, Earthscan Publications Ltd., 1988.

p. 127 **"On one recent trip..."** Jason Clay interview on CBC "Survival."

p. 128 **"So it's a way to revegetate..."** Judith Gradwohl, unbroadcast portion interview CBC "Survival."

p. 129 **Plan to "rent" rain forests from Third World nations...** James Goldsmith quoted by Reuter, March 26, 1990.

p. 130 **"We must find a way..."** Jose Lutzenberger, an ecologist who became Brazil's Environment minister, is quoted by Mark Edwards, "Forest Ire," *The Guardian*, March 30, 1990.

p. 130 **"I believe that we ought..."** E. O. Wilson interview on CBC "Survival."

Six: That's the Price of Progress

p. 133 **"This is a sign of progress..."** Mostafa Tolba, unbroadcast portion of interview for CBC "Survival."

p. 133 **"As for my own country..."** address by his Excellency Mr. Maumoon Abdul Gayoom before the U.N. General Assembly on the Issues of Environment and Development, New York, Oct. 19, 1987.

p. 134 **Our carbon dioxide emissions...** Peter Passell, "Global Warming: China Perplex," in the Economic Scene, *The New York Times*, March 7, 1990.

p. 134 **"In virtually all ..."** Sir Crispin Tickell in the NERC Annual Lecture at the Royal Society, London, June 5, 1989.

p. 135 **For example the United States ...** Hilary F. French, "A Most Deadly Tirade," *Worldwatch*, Vol. 3, No. 4, July/Aug. 1990, p. 14.

p. 135 **"One of the big problems ..."** Paul Erlich interview on CBC "Survival."

p. 136 **It is estimated ...** Hilary F. French quoting Professors Andrews Blowers of Britain's Open University and Denis Smith of Trent Polytechnic in "A Most Deadly Tirade," *Worldwatch*, Vol. 3, No. 4, July/Aug. 1990, p. 12.

p. 136 **Remember Koko?** Andrew Lees interview on CBC

"Survival."

p. 137 **"The greening of the north..."** all Martin Khor quotations from interview with British science journalist Fred Pearce, London, 1989.

p. 138 **As the century enters...** Alan B. Durning, Chapter 8: Ending Poverty; Worldwatch *State of the World 1990*, New York/London, W. W. Norton & Co., 1990, p. 135.

p. 139 **U.N. Children's Fund...** Ibid., p. 139.

p. 139 **Land is stripped of its trees...** Peter H. Raven of the Missouri Botanical Garden.

p. 139 **But between 1980 and 1987...** Ibid., p. 143.

p. 139 **"What people do not realize..."** Stephen Lewis unbroadcast portion of interview for CBC "Survival."

p. 140 **"Lifeboat ethics"...** Garrett Hardin in *Bioscience*, Vol. 24, 1974, p. 561.

p. 140 **"For a long time..."** Paul Ehrlich interview on CBC "Survival."

p. 141 **"We have no right..."** Michael Oppenheimer interview on CBC "Survival."

p. 142 **"CFCs are indispensable..."** Liu Ming Pu quoted by Samantha McArthur, "Campaign to save ozone must convince reluctant nations," Reuter, March 7, 1989.

p. 143 **At that level...** William K. Stevens quoting Michael Oppenheimer, "Ecological threats, rich-poor tensions," *The New York Times*, March 26, 1989.

p. 144 **"A total investment of $10 billion..."** Jose Goldemberg, Johansson, Reddy and Williams, *Energy for Development*, Washington, World Resources Institute, 1987, p. 34.

p. 144 **"Would require building..."** Ibid., p. 5.

p. 145 **"Once a $2 billion..."** Ibid., pp. 66–67.

p. 146 **"When you set up a scheme..."** Barry Commoner, author of *Making Peace With the Planet*, New York, Pantheon Books, 1990, personal communication.

p. 146 **A global carbon tax...** Chris Flavin proposals found in "Slowing Global Warming," Chapter 2, Worldwatch, *State of the World 1990*, New York/London, W. W. Norton and Co., 1990, pp. 27–29.

p. 146 **Canada could make a net cash saving...** Craig McInnes, "Supporting energy cuts would save Canadians $100 billion, report says," report by DPA Group Inc. for federal provincial Energy ministers; *The Globe and Mail*, Aug. 24, 1989.

p. 147 **We shouldn't forget the rich world's...** Noel Brown in unbroadcast interviews of UK TV program *Climate and Man*, 1989.

p. 147 **"We have no other choice..."** Jessica Tuchman-Matthews interview on CBC "Survival."

p. 148 **The first thing that the rich...** Paul Ehrlich in unbroadcast portion of interview for CBC "Survival."

p. 148 **"What we know now..."** Jessica Tuchman-Matthews interview on CBC "Survival."

p. 149 **"People have to begin to understand..."** Robert Ornstein unbroadcast portion of interview for CBC "Survival."

p. 149 **"I would like to see my children grow..."** Mahfuzal Haque interview on CBC "Survival."

Seven: Growth Is Progress

p. 152 **"Japan is now the number-one..."** Christian Blanckaert as quoted by Woody Hochswender, "Made in Japan: No fashion appeal at home," *The New York Times*, April 22, 1990.

p. 152 **Hardly a week goes by...** Rita Reif, "Rewriting Auction Records," *The New York Times*, Jan. 25, 1990; Steven R. Weisman, "2 top paintings go to Japan, but not all are happy there," *The New York Times*, May 19, 1990.

p. 152 **It is a national spending...** Clayton Naff, "Entertaining in Japan is costly, grueling, essential," *The Los Angeles Times*, Feb. 12, 1990; Thomas Walkom, "U.S. teacher forced to scavenge Tokyo," *The Globe and Mail*, March 8, 1988.

p. 153 **Take gold: Japan produces...** Andrew Tanzer, "Enjoy! Enjoy!," *Forbes*, Jan. 25, 1988, pp. 36–37.

p. 153 **The hunger for gold...** Karl Schoenberger, "Money boom; For Japan, gilded age of riches," *The Los Angeles Times*, Jan. 30, 1989; Fred Hiatt and Margaret Shapiro, "Japan's wealth creating conflict and self-doubt in the society," *The Washington Post*, Feb. 11, 1990.

p. 154 **A damning report...** François Nectoux and Yoichi Kuroda, "Timber From the South Seas; An Analysis of Japan's Tropical Timber Trade and Its Environmental Impact," World Wildlife Fund, March 1989.

p. 154 **"My impression is that..."** Sharon Begley, Hideko Tanayama, Mary Hager, "The World's Eco-Outlaw: critics take aim at Japan," *Newsweek*, May 1, 1989, p. 70.

p. 154 **80 tons of poisonous mercury dumped...** Stephan Schwartzman, Environmental Defense Fund, Washington, unbroadcast portion of interview for CBC "Survival."

p. 154 **"When you're talking..."** Jeanette Hemley, unbroadcast portion of interview for CBC "Survival."

p. 154 **"It's like saying it's cheaper..."** E. O. Wilson, unbroadcast portion of interview for CBC "Survival."

p. 155 **Description of Japan's international forestry** activities found in: Sharon Begley, Hideko Tanayama, Mary Hager, "The World's Eco-Outlaw," *Newsweek*, May 1, 1989, p. 70; Jeanette Hemley wwf; unbroadcast portion of interview for CBC "Survival", recorded May 31, 1989; Andrew Nikiforuk, "An Island's Hunger," in Habitat, *Equinox*, Nov./Dec. 1989, p. 165.

p. 155 **In April 1990, a young Swiss ...** interview with Bruno Manzer on CBC Radio "Quirks & Quarks," June 23, 1990.

p. 156 **Japanese had turned to the Soviets ...** Peter McGill, "Greener Japan sets mammoth task for Russians," *The Observer*, Dec. 31, 1989.

p. 156 **The minke is not ...** Sharon Begley, Hideko Takayama, Mary Hager, "The World's Eco-Outlaw," *Newsweek*, May 1, 1989, p. 70.

p. 156 **In a report ...** Dolphin killings, AP, London, June 29, 1990.

p. 157 **"I think we should be ..."** Ichiji Ishii, unbroadcast portion of interview for CBC "Survival."

p. 158 **"Japan is just one player ..."** Jeanette Hemley, unbroadcast portion of interview for CBC "Survival."

p. 158–159 **"Economics' benefits ..."** William Nordhaus, personal communication.

p. 160 **Anyone reading business ...** Lester R. Brown, editorial, *Worldwatch*, Vol. 3, No. 4, July/Aug. 1990, p. 2.

p. 160 **"The trouble is ..."** Paul Ehrlich interview on CBC "Survival."

p. 161 **Continuing to study economics...** Herman Daly interview on CBC "Survival."

p. 161 **"Economists are the only major ..."** Paul Ehrlich interview on CBC "Survival."

p. 161 **"For all its environmental damage..."** Bill Rees interview on CBC "Survival."

p. 162 **"To the millions of species..."** John Livingston: "The forest that once stood here and the water course that ran through it were homes for thousands of different species. The place was already fully developed.... The idea that undersea drilling is "development" of the sea bed presumably means "improvement" of the sea bed. In Darwinean terms every development is progress. Not to 'develop' would be to deny nature its sacred destiny in the human service," "A Planet for the Taking," Show 3, *Subdue the Earth*, 1985.

p. 163 **"I contribute $25 a year..."** Kenneth Boulding quoted by David Suzuki in *Inventing the Future*, Toronto, Stoddart, 1989, p. 113.

p. 164 **"We need to do..."** John C. Ryan, Economics of Scale, Worldwatch, Jan./Feb. 1990, p. 38.

p. 164 **One of the most serious...** Paul Ehrlich interview on CBC "Survival."

p. 164 **"The most powerful weapon..."** David Pearce quoted by Charles Clover, "This man is working out the bill for saving the planet. And he wants you to pay it," *Daily Telegraph*, Aug. 16, 1989.

p. 164 **"Quickly becomes nonsense..."** "How can polluters be made to pay?" *Nature*, Vol. 340, Aug. 24, 1989, p. 579.

p. 165 **"The only way..."** David Pearce, "Economists Befriend the Earth," *New Scientist*, Nov. 19, 1988, p. 34.

p. 165 **The idea that the economy...** Bill Rees unbroadcast portion of interview for CBC "Survival."

p. 167 **"I think we should..."** Jay Forrester, unbroadcast portion of interview for CBC "Survival."

p. 167 **The story of the havoc wreaked** by the Soviet plan for cotton in the Aral sea area is described by: Marjorie Sun, "Environmental Awakening in the Soviet Union," News and Comment Section, *Science*, Vol. 241, Aug. 26, 1988, pp. 1033–1035; Christian Tyler, "Worse than Chernobyl," *The Financial Times* (London), July 29/30, 1989.

p. 169 **"Growthmania"...** Herman E. Daly, "The Steady-State Economy: Alternative to Growthamania," Population-Environment Balance, April 1987.

p. 169 **"The way you've got..."** Paul Ehrlich interview on CBC "Survival."

p. 170 **"Since humans can't..."** Herman E. Daly, "Population and Development Review," Hoover Institution Conference, Spring 1990.

p. 170 **"There is no reason..."** Julian Simon quoted by David Suzuki in *Inventing the Future*, p. 115.

p. 170 **"The supposed scares don't exist..."** Julian Simon, personal communication.

p. 171 **"It's part of a general..."** Herman Daly, personal communication.

p. 171 **"If the diameter of the earth..."** C.H., "News and Comments," *Science*, June 17, 1988, p. 1611.

p. 172 **"We're not running out..."** Milton Friedman, personal communication.

p. 172 **"Our problems do not seem..."** Jay Forrester, unbroadcast portion of interview for CBC "Survival."

p. 172 **Miracle chemicals resulted...** Worldwatch Institute, *State of the World 1989*, New York/London, W. W. Norton & Co., 1989, p. 42.

p. 172 **They talked about a world...** Reuter, "Food surpluses seen as problem," *The Globe and Mail*, Dec. 9, 1986.

p. 173 **"Technological advances..."** Lester Brown quoted by AP, "Environmental disaster looming, institute says," *The Globe and Mail*, March 22, 1989.

p. 173 **From mid-century on...** Lester Brown, "Reexamining the World Food Prospect," Worldwatch Institute, *State of the World 1989*, New York/London, W. W Norton & Co., 1989, p. 54.

p. 173 **In the United States...** Sandra Postel, "Controlling Toxic Chemicals," Worldwatch Institute, *State of the World 1988*, New York/London, W. W. Norton & Co., 1988, p. 122.

p. 174 **If present farming practices...** Michael J. Dover and Lee M. Talbot, "Feeding the Earth, An Agroecological Solution," *Technology Review*, Feb./March 1988, pp. 27–35.

p. 174 **"We're forestalling..."** Jay Forrester unbroadcast portion of interview for CBC "Survival."

p. 174 **Pesticides are responsible...** Sandra Postel, "Controlling Toxic Chemicals," Worldwatch Institute, *State of the World 1988*, New York/London, W. W. Norton & Co., 1988, p. 122.

p. 174 **"No scientists believe..."** Paul Ehrlich interview on CBC "Survival."

p. 175 **"Wherever possible, we believe..."** George Bush speech to IPCC quoted by Michael Weisskopf "Bush pledges research on global warming," *The Washington Post*, Feb. 6, 1990.

p. 175 **"Faceless bureaucrats..."** John Sununu quoted by Bob Hepburn, "Bush rejects speedy action on curbing global warming," *Toronto Star*, Feb. 6, 1990.

p. 175 **"It is life itself..."** Margaret Thatcher (U.N. speech) quoted by Martin Walker, "Thatcher in call to save environment," *The Guardian*, Nov. 9, 1989.

p. 176 **"Among the serious problems..."** Brian Mulroney, opening address, conference on "The Changing Atmosphere, Implications for Global Security," Toronto, June 27–30, 1988 (UMO No. 710).

p. 176 **"Common sense tells us..."** Brian Mulroney (World energy conference 1989), quoted by Penny MacRae, "Energy, Budget," Montreal, CP, Aug. 29, 1989.

p. 177 **"If we don't move now..."** Lucien Bouchard interview on CBC "Survival."

p. 177 **"In view of the social..."** Jake Epp quoted by Shawn

McCarthy, "Energy ministers hedge on pollution cut," *Toronto Star*, Aug. 29, 1989.

p. 178 **It is telling that...** Anne McIlroy, "Canada brings home an F on report card of environment record," *The Gazette*, Montreal, July 8, 1990.

p. 178 **"I think the idea..."** Jay Forrester interview on CBC "Survival."

p. 179 **The system is just using...** Bill Rees interview on CBC "Survival."

p. 179 **Meeting the needs...** The World Commission on Environment and Development, *Our Common Future*, Oxford, Oxford University Press, 1987, p. 8.

p. 179 **"More rapid economic..."** Ibid., p. 89.

p. 179 **"During the most favourable..."** Ted Trainer, "A Rejection of the Brundtland Report," unpublished paper, p. 9.

p. 179 **Just isn't in the "ecological cards..."** Bill Rees interview on CBC "Survival."

p. 180 **"Away from the consumption..."** Herman Daly, personal communication.

p. 180 **So the goal of development...** Ted Trainer, "A Rejection of the Brundtland Report," unpublished paper.

p. 180 **What we've got right now...** Ralph Nader interview on CBC "Survival."

Eight: There at Our Disposal

p. 184 **Garbage statistics** are found in: Cynthia Pollock Shea, "Mining Urban Wastes, Potential for Recycling," Worldwatch Paper 76, Worldwatch Institute, April 1987; Cynthia Pollock Shea interview on CBC "Survival"; Alan Durning, "Ecology Starts at Home," Worldwatch, March/April, 1990; "Buried Alive," *Newsweek*, Nov. 27, 1989.

p. 185 **No place on Earth...** "Three-nation team plans Everest climb to clean up debris," AP, *Toronto Star*, Feb. 28, 1990.

p. 185 **Residents of a small Nova Scotia ...** Steve MacLeod, "Deadly debris haunts the sea," *The Ottawa Citizen*, June 29, 1989.

p. 186 **We have created a world...** Nancy J. White, "Sorting through the diaper dilemma," *Toronto Star*, Sept. 7, 1989.

p. 186 **"Our enormously productive..."** quoted by Janet Marinelli, "The Disposable Decades," *Garbage*, Sept./Oct. 1989, pp. 33–34.

p. 187 **"If you were to..."** Stuart Ewen as quoted by Woody Hockswender, "The green movement in the fashion world," *The New York Times*, March 25, 1990.

p. 187 **"It was disposability..."** Graham Decarie interview on CBC "Survival."

p. 187 **In Ontario alone, three-fifths of a hectare...** Robert A. Flemington, president of Ontario Multi-Material Recycling Inc., to a waste management conference in Toronto, "Just what was said," *The Globe and Mail*, July 10, 1989.

p. 188 **A description of a "state of the art" incinerator** — In his new book, *Making Peace with the Planet*, New York, Pantheon Books, 1990. Dr. Barry Commoner of the Centre for the Biology of Natural Systems at Queens College in Brooklyn writes: "According to Dr. Peter Montague of the Environmental Research Foundation in Princeton, N.J...." p. 119.

p. 188 **An additional 64 have been blocked...** Melinda Beck, with Mary Hager, Patricia King, Sue Hutchison, Kate Robins, Jeanne Gordon, "Buried Alive," *Newsweek*, Nov. 27, 1989, pp. 66–76.

p. 188 **Recycling potential** described in: Cynthia Pollock Shea, "Mining Urban Wastes: The Potential for Recycling," Worldwatch Paper 76, Worldwatch Institute, April 1987; Merilyn Mohr, "Burning Question," *Harrowsmith*, Mar/April 1988, pp. 43–54; "A cleaner environment, where to invest," *Changing Times*, Feb. 1990, pp. 32–39; Barry Commoner; interview on CBC "Survival."

p. 189 **"Recycling is just..."** Richard Gilbert interview on CBC "Survival."

p. 189 **Standardizing glass containers...** Cynthia Pollock Shea, "Mining Urban Wastes: the Potential for Recycling," Worldwatch Paper 76, Worldwatch Institute, April 1987.

p. 190 **"Half of the waste..."** Richard Gilbert interview on CBC "Survival."

p. 190 **Statistics on packaging** can be found in: Cynthia Pollock Shea, "Mining Urban Wastes, the Potential for Recycling," Worldwatch Paper 76, Worldwatch Institute, April 1987; Lynn Scarlett, "Pay for What You Throw Away," *USA Today*, Nov. 17, 1989; Melinda Beck, with Mary Hager, Patricia King, Sue Hutchison, Kate Robins, Jeanne Gordon, "Buried Alive," *Newsweek*, Nov. 27, 1989, pp. 66–76; John Holusha, "Why a squeezable bottle is under attack," *The New York Times*, Dec. 6, 1989; Ken MacQueen, "Targets set for reducing packaging," *The Gazette*, March 21, 1990.

p. 191 **In an EPA ranking...** Janet Marinelli, "Garbage at the Grocery," *Garbage*, Sept./Oct. 89, pp. 34–39.

p. 191 **"We're just using..."** Richard Gilbert unbroadcast portion of interview for CBC "Survival."

p. 192 **"The whole of the advertising..."** Bill Rees interview on CBC "Survival."

p. 192 **3SC Monitor study** described by Marina Strauss, "Young pleasure seekers born to shop," *The Globe and Mail*, Jan. 27, 1989.

p. 193 **"You know, I can see..."** Richard Gilbert interview on CBC "Survival."

p. 193 **"Why is it that..."** Al Gore interview on CBC "Survival."

p. 194 **"Deep forebodings that..."** Thomas Berry interview on CBC "Quirks & Quarks," broadcast April 21, 1990.

p. 194 **"Massive effort at every level..."** Bill Rees interview on CBC "Survival."

p. 195 **Statistics on the automobile** found in: Michael Renner, "Rethinking the Role of the Automobile," Worldwatch Paper 84, June 1988, p. 5; Michael Renner, "Car Sick," *Worldwatch* Nov./Dec. 1988 p. 36; Lester Brown, Christopher Flavin, Sandra Postel, "No Time to Waste," *Worldwatch* Jan./Feb., 1989; Kilaparti Ramakrishna, Woods Hole Research Institute, personal communication.

p. 195 **"If there is anything..."** Richard Gilbert unbroadcast portion of interview for CBC "Survival."

p. 195 **In 1916 James Doolittle described...** James Rood Doolittle, *The Romance of the Automobile Industry*, New York, Klebold Press, 1916, pp. v, ix, 441–42, cited in "A Runaway Match," by Julian Smith, in *The Automobile and American Culture*, Ann Arbor, University of Michigan Press, David L. Lewis and Laurence Goldstein, eds., 1983, p. 188.

p. 196 **You can trace the...** Julian Smith, "A Runaway Match," in *The Automobile and American Culture*, Ann Arbor, University of Michigan Press, David L. Lewis and Laurence Goldstein, eds. 1983, p. 182.

p. 197 **Governments in both countries ...** Michael Renner, "Rethinking the Role of the Automobile," Worldwatch Paper 84, June 1988, p. 6, p. 13.

p. 197 **Statistics on fuel consumption** in Third World from Richard Sandbrooke, president International Institute for Environment and Development Policy, London, England, interview for CBC "Survival"; David Bellamy and Brendan Quayle, *Turning the Tide*, London, William Collins and Son, 1986, p. 62.

p. 198 **Automobiles contribution of CFCs** from Janet Marinelli, "Mobile Air Conditioners," *Garbage*, Nov./Dec. 1989; p. 30; Irving Mintzer, Energy Policy analyst University of Maryland, "What on Earth Are We Doing?" a paper delivered at a conference on the environment, the 58th annual Couchiching Conference, Aug.

10–13, 1989.

p. 198 **The automobile's contribution to acid rain** discussed by David Israelson, "Group says Ottawa is bungling battle to reduce car smog," *Toronto Star*, Sept. 16, 1989.

p. 198 **The real cost of oil** discussed by: Gary Starr and Susan Bryer Starr, "The True Cost of Oil," *Earth Island Journal*, Summer 1988, pp. 22–23; "Oil spills likely, federal study says," *Toronto Star*, Dec 8, 1989; Mark Mardon, "Cars, Cars, Cars, Cars," *Sierra*, May/June 1989, pp. 22–28.

p. 199 **"We have to encourage ... "** Irving Mintzer, unbroadcast portion of interview for CBC "Survival."

p. 200 **Study for Toronto Transit Commission** described by Peter Howell, "Beef up ads to attract 'macho' riders, TTC Told," *Toronto Star*, Dec. 6, 1989.

p. 200 **Statistics on energy efficiency in automobile** found in: Arthur Rosenfeld, Energy Efficiency versus "draining America," Oversight hearing on: Oil development in ANWR and National Energy Policy, U.S. House of Representatives, March 31, 1988; Stephen Plotkin, "The Road to Fuel Efficiency," *Environment*, Vol. 31., No. 6, pp. 19–42.

p. 200 **Distances traveled by Americans** described by Robert Schaeffer, "Car Sick; Automobiles Ad Nauseam" *Greenpeace*, Vol. 15, No. 3, May/June 1990, pp.13–17.

p. 201 **"Large stretches of land ... "** Michael Renner, "Transportation Tomorrow," *The Futurist*, March-April, 1989, pp. 14–20.

p. 202 **Urban traffic problems** discussed in: "Damn that traffic jam," from *The Economist*, reprinted in the *The Globe and Mail*, March 24, 1989; Robert Schaeffer; "Car Sick; Automobiles Ad Nauseam," *Greenpeace*, Vol. 15; No. 3, May/June 1990, pp. 13–17; Marcia Lowe, "The Bicycle: Vehicle for a Small Planet," Worldwatch Paper 90, Worldwatch Institute, Sept. 1989, p. 18.

p. 202 **"By the middle ... "** Jack Layton quoted by Glenn Cooly, "Car Crackdown," *Now*, Nov. 23–29, 1989, p. 8.

p. 202 **Attempts to deal with urban traffic** described in: Mark Mardon, "Cars, Cars, Cars, Cars," *Sierra*, May/June 1989, pp. 22–28; "Damn that traffic jam," from *The Economist*, published in *The Globe and Mail*, March 24, 1989; "City sets goal to reduce carbon emission levels," *The Globe and Mail*, Jan. 31, 1990.

p. 203 **"Suppose a substantial tax ... "** David Crane, "Just wait until we have to pay cost of environment problems," *Toronto Star*, Feb. 7, 1990.

p. 203 **Every month in the L.A. basin ...** Larry Berg, interview on CBC "Survival."

p. 203   **Descriptions of the L.A. 20-year plan** found in: Alan Weisman, "L.A. fights for breath," *The New York Times Magazine*, July 30, 1989, p. 15–49; "California to require anti-smog computers for cars," Los Angeles, Reuter, Sept. 15, 1989; Ian Anderson, "Smog-Bound Los Angeles to Ban Petrol-Driven Cars," *New Scientist*, April 1, 1989, p. 21.

p. 205   **Poll on American views on mass transit** quoted in "Most Would Pay More for Cleaner Environment," *USA Today* poll, April 13, 1990.

p. 205   **More efficient vehicles** described in: David Bellamy and Brendan Quayle, *Turning the Tide*, Great Britain, William Collins and Sons, 1986, p. 59; Marcia Lowe, "The Bicycle: Vehicle for a Small Planet," Worldwatch Paper 90, Worldwatch Institute, Sept. 1989, p. 19.

p. 206   **"Our surveys indicate..."** Environics Research Group Ltd., "Environics Looks to the '90s," material drawn from syndicated studies: The 3SC Monitor of Social Change, the Environmental Monitor, Focus Canada, Focus Ontario, Homes National and Metro Poll.

Nine: Buying Time

p. 209   **"The inhabitants of planet Earth..."** Wallace Broecker as quoted by Irving Mintzer in the foreword of "A matter of degrees: The Potential for Controlling the Greenhouse Effect," World Resources Institute Report #5, April, 1987.

p. 209   **In recent months...** James R. Udell, "Turning Down the Heat," *Sierra*, July/Aug. 1989, p. 26–33.

p. 210   **We're not going to...** Stephen Lewis, interview on CBC "Survival."

p. 210   **The U.N. Intergovernmental Panel on Climate Change...** Paul Brown, "UN warns of global time-bomb," *The Guardian*, May 22, 1990; Craig R. Whitney, "UN warning on warming: Cut emissions 60% now, or else," *International Herald Tribune*, May 26–27, 1990.

p. 210   **"To virtually eliminate dependence..."** Stephen Lewis interview on CBC "Survival."

p. 211   **Statistics on North American energy use** from David Bellamy and Brendan Quayle, *Turning the Tide*, Great Britain, William Collins and Sons, 1986, p. 61; Dr. James Duke U.S. Dept. of Agriculture, "A Green World Instead Of A Greenhouse," *Earth Island Journal*, Summer 1988, pp. 29–31.

p. 211   **A frighteningly large gap...** Christopher Flavin, Chapter

2, Worldwatch *State of the World Report 1990*, New York/London, W. W. Norton & Co., 1990, p. 21; David Young, "Greenhouse gases to go up a third in 15 years," *The Times* (London), Nov. 17, 1989; Dennis Bueckert, CP energy story, Ottawa, Oct. 19, 1989.

p. 212 **"Do we have to destroy..."** James Watkins quoted in "More Research needed," *Nature*, Vol. 343, Feb. 1990, p. 684.

p. 212 **"Canada will not support..."** William Walker, "Minister denies U.S. deal to stall reduction of gases," *Toronto Star*, May 11, 1990.

p. 212 **"What they're doing is the height..."** Michael Oppenheimer interview with CBC "Quirks & Quarks," May 10, 1990.

p. 213 **"Our leaders have to take..."** Doug Scott interview on CBC "Survival."

p. 213 **"The important rethinking..."** Lester Brown interview on CBC "Survival."

p. 213 **"Carbon dioxide is the exhaling..."** Al Gore interview on CBC "Survival."

p. 214 **"We cannot have all..."** Doug Scott interview on CBC "Survival."

p. 214 **"It's a challenge to..."** Stephen Lewis interview on CBC "Survival."

p. 214 **"If you break the greenhouse..."** Stuart Boyle interview on CBC "Survival."

p. 215 **"We have the possibility..."** George Woodwell interview on CBC "Survival."

p. 215 **According to Worldwatch energy estimates...** Lester Brown "Toward A Climate-Sensitive Energy Policy," *Worldwatch*, Jan./Feb. 1989, p. 12.

p. 215–216 **The story of Osage, Iowa,** from: Wes Birdsall, "Economic Development and Changes in the Environment through Demand Side Management," 1989 speech; Ken Swenson, personal communication; "The Good News: Osage, Iowa, Counts Kilowatts," *Time*, Jan. 2, 1989, p. 37; Craig Canine, "Generating Negawatts," *Harrowsmith*, March/April, 1989, pp. 42–49; Richard Woodruff, personal communication.

p. 217 **Analysts at the American Council...** Christopher Flavin, "Slowing Global Warming: A Worldwide Strategy," Worldwatch Paper 91; Oct. 1989, p. 44.

p. 217 **Each compact bulb...** James R. Udall, "Turning Down the Heat," *Sierra* July/Aug. 1989, pp. 26–33.

p. 217 **U.S. utilities alone...** Matthew L. Wald, "What it would take to fight the greenhouse effect," *The New York Times*, Aug. 28,

1990.

p. 218 **"It's worth more than..."** Irving Mintzer unbroadcast portion of interview for CBC "Survival."

p. 218 **"The tube of caulk..."** Ted Flannigan, personal communication.

p. 218 **Some superinsulated homes...** John H. Gibbons, Peter Blair and Holly Gwin, "Strategies for Energy Use," *Scientific American*, Sept. 1989, p. 140.

p. 219 **"If you take superwindows..."** Arthur Rosenfeld, personal communication.

p. 219 **Carbon dioxide output...** George Woodwell interview on CBC "Survival."

p. 220 **Nuclear energy figures from:** "Nuking the Greenhouse," *New Scientist*, Nov. 5, 1988, p. 20; Lester Brown, "Toward A Climate-Sensitive Energy Policy," *Worldwatch*, Jan./Feb. 1989; p. 12; Charles Komanoff, Komanoff Energy Assoc., "Greenhouse Effect Amelioration — Efficiency vs. Nuclear," *Energy Newsletter*, Aug. 24, 1988; Energy Policy, Rocky Mountain Institute, Dec. 1988; Christopher Flavin, Chapter 2, "Slowing Global Warming," Worldwatch *State of the World 1990*, New York/ London, W. W. Norton and Co., p. 23.

p. 221–222 **Alternative energies** discussed in: Christopher Flavin, p. 27; *A Report by the International Institute for Environment and Development and the World Resources Institute*, New York, Basic Books Inc., 1987, pp. 93–108.

p. 223 **Solar energy** discussed in: Christopher Flavin, p. 25; Matthew L. Wald, "Business technology; A Cheaper Road to Solar Power: It's Done With Mirrors," *The New York Times*, Oct. 4, 1989, p. 30; Dr. H. M. Hubbard, director Solar Energy Research Institute, testimony before U.S. House of Representatives Committee on Science, Space and Technology Sub-Committee on Energy Research and development, June 30, 1988; *A Report by the International Institute for Environment and Development and the World Resources Institute*, New York, Basic Books Inc., 1988–89, Chapter 7, Energy, page 114. Lester Brown, Christopher Flavin, and Sandra Postel, "Picturing a Sustainable Society," Worldwatch *State of the World 1990*, New York/London, 1990, W. W. Norton & Co., Chapter 10, pp. 175–179.

p. 225 **Solar funding statistics** found in: Martin Mittelstaedt, "Solar power's appeal grows as cost falls," *The Globe and Mail*, Dec. 5, 1989; *Now*, May 5–11, 1988.

p. 225–226 **Tree planting** discussed in: "Reforesting the Earth," *Worldwatch*, Jan./Feb. 1989, pp. 13–15; Greg Marland of the Oak

Ridge National Laboratory, "The Prospect of Solving the CO_2 Problem Through Global Reforestation," paper delivered at American Association for Advancement of Science meeting, Feb. 12, 1988, p. 59; Christopher Flavin, "Slowing Global Warming," pp. 17–38.

p. 227 **"I don't like to be pessimistic..."** George Woodwell unbroadcast portion of interview for CBC "Survival."

p. 229 **What's at stake at minimum...** Stephen Lewis, unbroadcast portion of interview for CBC "Survival."

p. 229 **"We will flirt with..."** Arthur Rosenfeld interview with Jay Ingram, CBC "Quirks & Quarks," broadcast April 21, 1990.

p. 231 **"We are in an era..."** Robert Ornstein unbroadcast portion of interview for CBC "Survival."

Ten: A Sense of Place

p. 233–234 **Story of disappearing amphibians** found in: Ron Cowen, "Tales From the Froglog and Others," *Science News*, Vol. 137, Mar. 10, 1990, p. 158.

p. 235 **"In biological terms..."** Timothy C. Weiskel, "While Angels Weep... Doing Theology on a Small Planet," *Harvard Divinity Bulletin*, Fall 1989, Vol. X1X, No. 3.

p. 236 **"Put the Bible on the shelf..."** Father Thomas Berry interview on CBC "Quirks & Quarks," broadcast April 21, 1990.

INDEX